PRINCIPLES OF SUSTAINABLE AQUACULTURE

A SYSTEMATIC APPROACH

Authors

Dr. Rekha Moun. She is a Senior Research Fellow at CIRB, Hisar, and achieved her doctorate in Zoology from Chaudhary Charan Singh Haryana Agricultural University, Hisar. Holding a postgraduate degree in Zoology from Maharishi Dayanand University, she has become a notable figure in the field. Dr. Moun has published many articles on aquaculture. As the editor of the book "Principles of Sustainable Aquaculture: A Systematic Approach " she has contributed in creating a valuable resource for enthusiasts and researchers alike, exploring fundamental principles in fish culture and management.

Mr. Shri Ram Yadav. He has been post graduated in M.F.Sc. (Fisheries Resource Management) in 2022 from CCSHAU, Hisar (Haryana). He has published one text book and several research publications on the fisheries sector in reputed journals. Mr. Yadav has received Best Thesis Award (PG), Young Research Award, at international conference.He is a member of ISAHRD, Chandigarh. His fields of interest are aquaculture and fish disease & management.

Dr. Ravikant works as an Assistant Professor (Stage III) in the Department of Zoology and Associated faculty in the Departments of Fisheries Resource Management and Aquaculture, College of Fisheries Science, CCS HAU, Hisar (Haryana). He has over 12 years of experience in teaching, research and extension. He was interested in building entrepreneurship in shrimp and fish cultures and involved himself in various research activities. He has more than 100 publications to his credit including six books, and various research papers in leading international and national journals.

Dr. Rajender Kumar Gupta, Professor & Head, Department of Zoology and Associated faculty in the Departments of Fisheries Resource Management and Aquaculture in College of Fisheries Science, CCS HAU, Hisar (Haryana). He has over 30 years of rich experience in teaching, research and extension. Dr. Gupta visited more than 15 countries and chaired various International academic responsibilities. He is always actively contributed in building entrepreneurship in aquaculture related assignments. He has authored more than six books and over a hundred research papers/ articles.

PRINCIPLES OF SUSTAINABLE AQUACULTURE

A SYSTEMATIC APPROACH

– Authors –

Rekha Moun

Shri Ram Yadav

Ravikant

R K Gupta

ASTRAL

Publishing Authors Across the Globe

2025

Daya Publishing House®

A Division of

Astral International Pvt. Ltd.

New Delhi – 110 002

ISBN: 978-93-5919-473-8

Published by : **Daya Publishing House®**
A Division of
Astral International Pvt. Ltd.
– ISO 9001:2015 Certified Company –
4736/23, Ansari Road, Darya Ganj
New Delhi-110 002
Ph. 011-43549197, 8130496929
e-mail: info@astralint.com
Website: www.astralint.com

Printed at : **Replika Press Pvt. Ltd.**

'This dedication goes out to the farmers, the unsung heroes who nourish our bodies and souls'

PREFACE

This book, Principles of Sustainable Aquaculture: A Systematic Approach, is your deep dive into this exciting field. As the global population continues to rise, the demand for healthy, nutritious seafood is placing an ever-increasing burden on our water bodies. Aquaculture, the farming of aquatic plants and animals, offers a promising solution to meet this growing need. However, ensuring this growth is sustainable – environmentally, economically, and socially – is paramount.

This book is designed to equip students, researchers, and farmers with the essential knowledge and tools to navigate this critical field. This text emphasizes a science-based understanding of aquaculture principles, for stakeholders who can understand this dynamic field, as well as professionals looking to expand their knowledge and implement sustainable strategies.

In order to make this book more interesting, each chapter is written in simple language with suitable pictures. The journey of writing and editing this collection has been both enlightening and humbling. Each chapter represents a unique perspective and a significant addition to the existing body of knowledge. Our goal has been to create a cohesive and valuable resource that will inspire further academics, research and discussion.

We would like to express our sincere appreciation to our teachers for their invaluable intellectual contributions and guidance throughout this endeavor. Additionally, we extend our heartfelt thanks to the dedicated editorial team for their tireless efforts in bringing this volume to fruition. Lastly, we extend our gratitude to the readers who will engage with the contents of this volume. The authors hope that this book will serve as a valuable resource, sparking new ideas, fostering dialogue, and promoting a deeper understanding of the subjects explored within. We are highly thankful to our university's Worthy Vice-chancellor for providing a platform to create the manuscript. Last but not least, we sincerely thank Dr. Neeraj Kumar, Dean, COBSH & COFS, and all faculty members, CCS Haryana Agricultural University, Hisar for their valuable suggestions and their kind support during the preparation of this book. The interest taken by the publisher in this book is also highly appreciated.

Rekha Moun
Shri Ram Yadav
Ravikant
R K Gupta

CONTENTS

CHAPTER 1

AQUACULTURE AND ITS ECONOMIC STATUS

Introduction

The controlled cultivation of aquatic species, aquaculture or aquafarming, is a rapidly rising industry in developing nations like China and India. It's not just a local trend. It is a global phenomenon reshaping the world's food production and trade landscape (FAO, 2020 and Tacon, 2022). Fish farming (Pisciculture), ornamental fish farming, oyster and shrimp farming, mariculture, and algaculture (including seaweed farming) are examples of specific types of aquaculture.

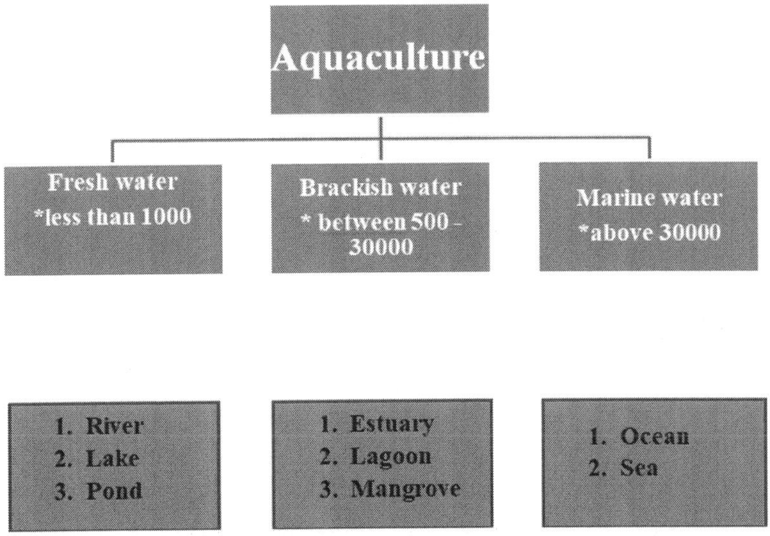

Fig. 1 Classification of aquaculture resources based on salinity level (Source: Authors)
***Salinity in parts per million (ppm)**

The three primary aquaculture subsectors are shellfish, freshwater finfish, and marine finfish. The marine finfish industry is the most significant commercially, with a production value of €1.8 billion in 2018. It is followed by the shellfish sector, which generated €1.3 billion, and the freshwater sector, which generated €1 billion (Sala *et al.*, 2015).

Over the past 50 years, the amount of fish consumed globally per person has almost doubled, and on an edible weight basis, it is now comparable to that of pig and fowl. For example, since 1990, China's per capita annual intake of aquatic food has more than tripled, rising from 11.2 kilograms to 38.0 kg in 2018. This trend has coincided with the expansion of aquaculture, with China presently accounting for roughly 35% of the world's consumption of aquatic food and 57% of the world's production of aquaculture. The average annual intake per person worldwide is 20.21 kilograms. Despite being one of the most dynamic industries in the world food chain, aquaculture is still remarkably underrepresented in most food policy literature. Growth in aquaculture production has matched and encouraged this rise in fish demand and will keep doing so.

In many cases, aquaculture provides a healthier and more environmentally friendly substitute for the consumption of red meat (Garlock *et al.*, 2022). Between 660 and 820 million people depend on the fishing and aquaculture industries for their livelihoods; these individuals make up 10 to 12 % of the global population (Taves, 2015). Asia leads the globe in aquaculture production, in 2020, it accounted for approximately 91.6% of the worldwide output, with China alone contributing 56.7% (FAO, 2022). China and India are the two dominant nations in Asia in terms of the global fishing sector. China is the primary producer, consumer, and trader of fish food products (Telegraph, 2021). India ranks third in the world for fish production, accounting for 7.96 per cent of total output worldwide. It also contributes more than 6.72% to the agricultural GDP and over 1.09% to the nation's total GDP (Sivagnanam, 2014). Between 2020 and 2021, the fisheries sector showed an exceptional 10.34% growth rate in fish output, with an average annual growth rate of 8% for the previous eight years.

In 2021–2022, the combined production of fisheries and aquaculture hit a record-breaking 214 million metric tonnes (MMT), of which 178 MMT were aquatic animals and 36 MMT were algae. An estimated 16.25 MMT of fish were produced overall in FY 2021–2022, with the inland sector contributing 12.12 MMT. India is a significant aquaculture fish producer as well. Since the nation's independence, the amount of fish produced there has increased steadily and continuously. Fifty-six per cent of the potential for marine fisheries and fifty-one per cent for inland fisheries have been utilised in FY 2020–21. India is the world's fifth-largest exporter of fish and fishery goods, boosting the country's reputation internationally. The industry employs several lakh people throughout the value chain in the nation, including 20 million primary fishermen and fish farmers, and generates more than Rs. 1.75 lakh crores in revenue from exports each year.

Table 1.1 Global contribution of pisciculture in India

Area	Contribution
Aquaculture	2nd rank
Fish Production	3rd rank
Indian GDP	1.09%
Agricultural GDP	5.5%
Global fish diversity	10%
Fish export	5th rank
Total production	214 million metric tonnes
Top most states	Andhra Pradesh, West Bengal, Gujarat, Kerala, Tamil Nadu

Source: Sinha & Damroy, 2024; FAO, 2020

Fisheries status in Haryana

After being divided in 1966 from the most impoverished area of Punjab, the state of Haryana has grown to become known as one of the wealthiest in the nation. To support the development of fisheries, the state possesses an abundance of natural and artificial lakes, reservoirs, micro-water sheds, canals, drains, and village ponds. Just 58 ha of pond water were stocked with 1.5 lakh fish seeds in 1966–1967, and only 600 tonnes of fish were produced annually. By FY 2022–2023, fish production will have increased to 21,2042.51 MT, with 19,100 ha of area under cultivation thanks to the stocking of 6917.36 lakh fish seeds. By covering a 25100.00 ha water area, it is planned to stock 5775 lakh Fish fingerlings to produce 228000.00 MT of fish during the 2023–2024 season. However, money has no colour or creed, as the saying goes. Though it was difficult to persuade them initially, a few ambitious farmers adopted fish culture in the early 1980s primarily due to its economic sustainability and profitability. In light of this, the development of fisheries has advanced significantly over the past 40 years, mainly supplying the Delhi market. The Directorate of Fisheries, Haryana, which began with 58 hectares of fish farms in 1966, has grown to include over 21,000 hectares of ponds throughout the State and has established a reputation for excellence in fisheries development throughout India. An approach with a focus is essential for exponential growth. According to a recent Haryana Fisheries Department report, fish growers earn a respectable 3.20 lakh net profit annually per hectare from their waterlogged property. Haryana has been proclaimed a "fish disease-free state" by the National Bureau of Fish Genetic Resources due to the State's outstanding success in fish breeding. Famous fish species like rohu, mrigal, catla, common carp, silver carp, and grass carp are farmed by fish farmers in Haryana in addition to shrimp and prawns. In addition, the Government of Haryana is investing 13.68 crore to construct a state-of-the-art decorative fish hatchery in the Jhajjar district to foster ornamental fish culture.

Standard fishing technology and commercialised fish

The technology involved in catching fish is the focus of the fishing industry. Some of the earliest fishing methods were hand-picking, spearing and harpooning, grappling and injuring, permanent and semipermanent obstacles, line fishing, and gear constructed of wood, twigs, and flexible branches. Larger, more potent fishing gear and boats with greater carrying capacities have supplanted these items. Crafts include gill nets, trap nets, trawls, purse seines, gill nets, seines traps, and trawlers. Fish hatchery in Jhajjar district that is ornamental, costing 13.68 crore. Freshwater fishes represent a wide range of fish specialised to inhabit freshwater habitats like rivers, lakes, ponds, and streams. Commercially viable freshwater fishes are those commonly cultivated and nurtured for diverse purposes such as sustenance, aquaculture, and the ornamental trade.

1. **Carp (*Cyprinus carpio*):** Throughout the world, various carp species, including standard, mirror, and grass carp, are extensively raised for food. They can withstand a variety of climatic conditions and develop quickly.
2. **Catfish (*Clarias* spp.)**: Because of their quick growth rates and strong market demand, catfish species like the African catfish (*Clarias gariepinus*) and Pangasius catfish (*Pangasianodon hypophthalmus*) are widely used in aquaculture.
3. **Trout (*Oncorhynchus* spp., Salmo spp.)**: Commercial trout farming occurs on freshwater farms, especially in areas with appropriate water conditions. It includes species like brown and rainbow trout. Their exquisite flavour and texture make them highly valued.
4. **Salmon (*Salmo salar, Oncorhynchus* spp.)**: Salmonids, such as the Atlantic and Pacific salmon species, are raised in freshwater hatcheries and then released into the ocean to continue growing. Their high market demand and nutritional benefits make them desirable.
5. **Tilapia (*Oreochromis* spp.)**: One of the most popular freshwater fish raised worldwide is tilapia. They are renowned for their quick development, environmental flexibility and mild meat flavour.
6. **Barramundi (*Lates calcarifer*)**: Barramundi, sometimes referred to as Asian sea bass, are raised in freshwater ponds and tanks for their mild flavour and firm white meat. Their high market demand and rapid growth make them attractive in aquaculture.
7. **Sturgeon (*Acipenser* spp.)**: Sturgeon species such as beluga, sevruga, and Russian sturgeon are cultured for their prized caviar. Sturgeon farming requires extensive, well-managed facilities due to these fish's slow growth and specific environmental requirements.
8. **Gourami (*Osphronemidae* spp.)**: In the ornamental fish trade, gouramis, including varieties like the gigantic gourami and pearl gourami, are highly sought-after. Because of the mild flavour of their flesh, they are also raised for food in some areas.

9. **Murrel (*Channa* spp.)**: In several Asian nations, murrel, often called snakehead fish, are raised for their meat and prized in specific regional specialities. They are suitable for aquaculture because they are resilient and tolerant of low oxygen levels.

10. Promoting and selling marine fish entails marketing species that thrive in saltwater environments, often for aquaculture, the aquarium trade, or consumption. These fish encompass a variety of species, from common aquarium pets to prized seafood delicacies. Here is a concise introduction along with examples of commercially significant marine water fish:

Aquarium Trade:

1. **Clownfish (*Amphiprion* spp.):** Clownfish, prized by aquarium enthusiasts worldwide for their vivid hues and harmonious coexistence with anemones, are highly sought-after pets.
2. **Tangs (*Zebrasoma* spp.):** These vibrant, herbivorous fish enhance the aesthetic appeal of saltwater aquariums and are highly valued for their capacity to consume algae.
3. **Blue Tang (*Paracanthurus hepatus*):** The blue tang, made famous by Finding Nemo, is a popular option for marine aquarium owners.

Seafood Consumption:

1. **Tuna (*Thunnus* spp.):** Many kinds of tuna, including yellowfin and bluefin, are highly valued for their tasty flesh and are economically significant to the fishing industry.
2. **Salmon (*Salmo* spp.):** Although certain species of salmon migrate between freshwater and saltwater, commercial salmon farms raise their fish in marine conditions to harvest their meat, which is rich in nutrients and flavour.
3. **Cod (*Gadus* spp.):** Cod is a common ingredient in many dishes and a significant contributor to commercial fisheries. It is found in cold, deep waters.

Ornamental Fish Trade:

1. **Moorish Idol (*Zanclus cornutus*):** The Moorish idol is a prized addition to marine aquariums. Its distinctive appearance is characterised by long, trailing dorsal fin extensions.
2. **Angelfish (Pomacanthidae family):** These graceful, vividly coloured fish are well-liked for marine aquariums because they lend a delicate touch to the underwater scenery.

Picture 1: A fish pond with hapas (Source: Authors)

Picture 2: Common fish pond (Source: Authors)

Government schemes for fish farming in India

The geographic distribution of aquaculture expansion, along with the species kinds, technologies, management techniques, and infrastructure employed in various regions, have all been significantly affected by government regulations (Garlock *et al.*, 2020). With over 650 species of fish, shellfish, aquatic plants, and algae farmed in 2020 in a range of freshwater, brackish, and marine habitats and traded extensively, the aquaculture industry is extensive and complicated. Many different types of aquaculture operations are found all over the world. Some examples include large brackish ponds of highly valued shrimp, on-land raceways for trout, intensive commercial ponds or cages of tilapia and pangasius (catfish) raised in freshwater systems, hanging lines of mussels and macroalgae (seaweed) in nearshore coastal zones, and marine netpen aquaculture, large submersible offshore cages, and land-based recirculating systems raising a variety of high-valued species (Short *et al.*, 2021). Public investments in infrastructure (e.g., ponds, waterways, roads, cages, and ports) and basic research and development (R&D, e.g., genetic improvements in fish, shellfish, and algae, research on fish nutrition and health) provide the foundations for aquaculture innovation and expansion in all producing countries. Innovation is the engine for aquaculture productivity growth and reduced input prices for producers and consumers, as emphasised by (Anderson *et al.*, 2019) in their review of key economic contributions related to aquaculture technology. The Indian government backs the country's fisheries sector and has introduced multiple schemes and initiatives to boost its development.

Pradhan Mantri Matsya Sampada Yojana (PMMSY): The Indian government has given the green light to the Pradhan Mantri Matsya Sampada Yojana scheme. Which aims to foster a 'Blue revolution' through sustainable and accountable growth in the fisheries sector. This initiative involves a total investment of Rs 20,050 crores and spans five years, encompassing all states and Union Territories (UT). The fisheries sector directly or indirectly sustains the livelihoods of approximately 28 million individuals in the country (Chandrappagari *et al.*, 2023).

Critical components of PMMSY include:

- **Infrastructure construction**: To improve post-harvest management and value addition, the plan strongly emphasises the building of fisheries-related infrastructure, such as fishing harbours, cold chain facilities, fish landing centres, and processing units.
- **Integrated strategy**: To ensure the comprehensive development of the fisheries sector, PMMSY uses an integrated approach by integrating different sub-schemes such as the National Fisheries Development Board (NFDB), National Fisheries Development Scheme (NFDS), and National Livestock Mission (NLM).
- **Employment generation**: Through skill development, training, and entrepreneurship development initiatives, PMMSY focuses on generating employment possibilities in the fisheries sector, especially for youth and women.
- **Insurance**: One of the PMMSY plan's subcomponents, insurance for these fishermen, falls under the beneficiary-oriented centrally sponsored scheme at Sl No. 14.1 of Annexure-II. Under this scheme, fishermen between the ages of 18 and 70 are eligible for coverage in the following ways:

  I. Rs. 5.00 lakh against total permanent incapacity or death

  II. Rs. 2.50 lakh for a lifetime partial handicap

  III. Rs. 25,000 to prevent unintentional hospital admission: By encouraging the formation of farmers' producer organisations (FPOs) and fish farmer producer organisations (FFPOs) and the development of market infrastructure to guarantee more favourable prices for their produce, PMMSY helps fishers and fish farmers establish market links.

Fish Pond Yojana is a government-backed programme offering Indian farmers financial support to expand freshwater aquaculture water bodies. It gives Indian farmers financial support and subsidies to build more bodies of water, including ponds and tanks, to raise fish.

Key features of the Fish Pond Yojana includes:

- **Subsidies**: This program provides financial assistance to farmers to build new fish ponds or enlarge existing ones. The subsidies mainly cover the ponds' building, excavation, and liner costs.
- **Technical Support**: To maximise fish output, the program offers farmers technical advice and assistance with pond design, pond management procedures, fish seed stocking, feeding strategies, disease control, and harvesting techniques.
- **Capacity Building**: To help farmers become more capable and equipped with the knowledge and skills necessary for successful fish farming, The Fish Pond Yojna conducts awareness campaigns, workshops, and training programs. The Fish Pond Yojna provides incentives for establishing new freshwater aquaculture bodies to increase fish productivity, boost farmer incomes, and bolster food security and rural livelihoods in India.

Applications of Aquaculture:

1. Rapid production growth has increased seafood's significance for ensuring nutrition and food security in several countries (Naylor *et al.*, 2023).
2. It offers a necessary means to meet growing food demand and address nutritional deficiencies (Golden *et al.*, 2021)
3. The consumption of aquatic foods improves human health in three ways. First, lower the risk of subsequent disease caused by micronutrient deficiencies (such as those involving vitamin A, calcium, and iron). Second, by serving as the primary source of the omega-3 long-chain polyunsaturated fatty acids docosahexaenoic acid (DHA) and eicosapentaenoic acid (EPA), which together may lower the risk of heart disease and improve brain and eye health. Third, by substituting the consumption of less healthful red and processed meats that may have adverse health effects (Rimm *et al.*, 2018).
4. Consuming aquatic food is essential for the development of prenatal neurocognition and adult cardiovascular health because it is high in bioavailable micronutrients, including zinc, calcium, and long-chain Ω-3 polyunsaturated fatty acids (Mohan *et al.*, 2021).
5. Over the previous three decades, aquaculture has more than doubled the amount of aquatic food available, rising from 13.1 million mt in 1990 to 82.1

million mt in 20181. Further expansion is likely guaranteed (Kobayashi *et al.*, 2015).

6. Well-managed aquaculture can contribute to environmental conservation by reducing pressure on wild fish stocks and habitats. Aquaculture helps alleviate the strain on marine ecosystems and supports biodiversity conservation by providing an alternative to wild-caught seafood. Furthermore, responsible aquaculture practices prioritise environmental sustainability, promoting ecosystem health and resilience.
7. Compared to traditional fishing methods, aquaculture can be more resource-efficient. Controlled environments allow for optimised feeding, reduced waste, and better management of water resources. Additionally, aquaculture systems can recycle nutrients and minimise environmental impact through integrated multi-trophic aquaculture (IMTA) practices.
8. Aquaculture boosts economic growth and produces jobs, especially in rural and coastal areas where job options may be scarce. By encouraging entrepreneurship and investment in aquaculture businesses, societies can gain higher living standards and more revenue.

References

Anderson, J. L., Asche, F. & Garlock, T. (2019). Economics of aquaculture policy and regulation. Annual Review of Resource Economics, 11, 101-123.

Chandrappagari, S., Ajith, S. & Olive, R. J. (2023). Implementation of the Group Accident Insurance Scheme (GAIS) under Pradhan Mantri Matsya Sampada Yojana (PMMSY)–Impact on fisheries sector across INDIA. http://dx.doi.org/10.2139/ssrn.4644996

Covering Death, P. G. & Policy, S. C. Pradhan Mantri Matsya Sampada Yojana (PMMSY). Policy, 6, 0-40.

FAO (2020). The State of World Fisheries and Aquaculture. Rome: FAO, United Nations.

FAO (2022). The State of world fisheries and aquaculture (2022). Towards blue transformation. FAO. https://doi.org/10.4060/cc0461en

Garlock, T., Asche, F., Anderson, J., Bjørndal, T., Kumar, G., Lorenzen, K. & Tveterås, R. (2020). A global blue revolution: aquaculture growth across regions, species, and countries. Reviews in Fisheries Science & Aquaculture, 28(1), 107-116.

Garlock, T., Asche, F., Anderson, J., Ceballos-Concha, A., Love, D. C., Osmundsen, T. C. & Pincinato, R. B. M. (2022). Aquaculture: The missing contributor in the food security agenda. Global Food Security, 32, 100620.

Golden, C. D., Koehn, J. Z., Shepon, A., Passarelli, S., Free, C. M., Viana, D. F. & Thilsted, S. H. (2021). Aquatic foods to nourish nations. Nature, 598(7880), 315-320.

Kobayashi, M., Msangi, S., Batka, M., Vannuccini, S., Dey, M. M. & Anderson, J. L. (2015). Fish to 2030: The role and opportunity for aquaculture. Aquaculture economics & management, 19(3), 282-300.

Mohan, D., Mente, A., Dehghan, M., Rangarajan, S., O Donnell, M., Hu, W. & PURE, O. (2021). Associations of fish consumption with risk of cardiovascular disease and mortality among individuals with or without vascular disease from 58 countries. JAMA Internal Medicine, 181(5), 631-649.

Naylor, R., Fang, S. & Fanzo, J. (2023). A global view of aquaculture policy. Food Policy, 116, 102422.

Rimm, E. B., Appel, L. J., Chiuve, S. E., Djoussé, L., Engler, M. B., Kris-Etherton, P. M & Lichtenstein, A. H. (2018). Seafood long-chain Ω-3 polyunsaturated fatty acids and cardiovascular disease: a science advisory from the American Heart Association. Circulation, 138(1), e35-e47.

Sala, A., Martin, P., Maynou, F. & Damalas, D. (2015). Scientific, Technical and Economic Committee for Fisheries (STECF)-Landing Obligation-Part 6 (Fisheries targeting demersal species in the Mediterranean Sea)(STECF-15-19).

Short, R. E., Gelcich, S., Little, D. C., Micheli, F., Allison, E. H., Basurto, X. & Zhang, W. (2021). Harnessing the diversity of small-scale actors is critical to the future of aquatic food systems. Nature Food, 2(9), 733-741.

Sinha, A., Gogoi, P. & Damroy, S. (2024). Small Indigenous Fish (SIF): Status and Contributions in Nutrition and Livelihood Security of India: A Review. Agricultural Reviews, 45(1).

Sivagnanam, K. J. (2014). State Agriculture Profile of Tamil Nadu–2011. Agro-Economic Research Centre, University of Madras, Chennai, Tamil Nadu, India, 1-98.

Tacon, A. G., Metian, M. & McNevin, A. A. (2022). Future feeds: suggested guidelines for sustainable development. Reviews Fisheries Science & Aquaculture, 30(2), 135–142.

Taves, J. R. (2015). Assessing vulnerability in households and communities involved in fisheries sector activities (Doctoral dissertation, Université d'Ottawa/ University of Ottawa).

Telegraph (2021). India ranks No. 2 in fish production, but China ranks No. 1 ten times bigger—the Telegraph Online. https://www.telegraphindia.com/india/india-ranks-no-2-in-fish-production-but-no-1-china-is-ten-times-bigger/cid/1578442

CHAPTER 2

TYPE OF FISH HATCHERIES

Various hatcheries have been created to enhance fish egg hatching. Their primary goal is to increase the hatching success rate. The different hatchery types include:

1. Earthen hatching pits

In the early days, a pioneering method of hatching was introduced in earthen hatching pits. These pits, with their standard dimension of 3 feet in length, 2 feet in width, and 1 foot in depth, revolutionized the hatching process. They were not just holes in the ground but meticulously designed structures. Several pits were dug in lines, their insides covered with mud. After filling them with water, the eggs gathered were deposited into these pits. Each pit could hold around 35,000 to 40,000 eggs to hatch. The hatching process typically occurs within 24 hours. The pits were interconnected and equipped with proper irrigation and drainage systems. Maintaining a constant flow of water was crucial as it facilitated adequate aeration and minimized waste accumulation, ultimately improving the survival rate of the hatchlings. On average, the hatching success rate in these pits ranged from 30% to 40% (Das, 2003).

The utilization of earthen hatching pits offers several advantages:

- Firstly, these depressions are highly suitable for incubating eggs in arid embankments. By utilizing expansive areas near dry embankments, it becomes feasible to dig earthy hollows, thus reducing the number of eggs needed for each depression.
- Secondly, the fresh rainwater that accumulates in the bunds can naturally enter the pits, providing a suitable environment for hatching. This natural water source facilitates hatching and ensures that the eggs receive the necessary moisture for successful incubation.

- Lastly, one of the significant benefits of employing earthen hatching pits is the low expenditure involved. The technology is inexpensive, making it a cost-effective option for egg hatchery operations. This affordability makes it accessible to a broader range of individuals and communities, promoting its adoption and use.

These hatching pits have some drawbacks, notably a high mortality rate due to temperature fluctuations during open-area hatching. Oxygen depletion is another issue, resulting in significant spawn mortality. Maintaining continuous water flow in the pits until spawn collection is crucial to prevent mortality due to insufficient water. The Chittagong type of hatching pits is similar to earthen pits but includes a cloth and mosquito net in each pit. The fabric is placed just above the pit's bottom, while the mosquito net is positioned above the cloth. Spawn passes through the net during hatching and collects on the fabric. After three days, the net containing eggshells and dead eggs is removed. The spawn is carefully removed once the yolk sac is fully absorbed (Dickson & Fanelli, 2013).

2. Chinese hatchery:

The Chinese method of breeding carp and hatching eggs relies on a continuous gravitational water flow. This approach offers numerous benefits, such as reduced construction and operational expenses, compared to alternative designs of similar productivity. In India, the Chinese hatchery system is now widely acknowledged for its ability to yield high-quality fish offspring. Comprising four key elements—an elevated water storage tank, a spawning/breeding pond, an incubation hatching pond, and a hatchling receiving pond—this system is purpose-built for fish breeding and incubation. Water for the hatchery is supplied through a regulated pipe connected to the overhead tank. The hatching cycle typically spans four days, followed by a four-day interval before it can be repeated.

The overhead water storage tank is designed to be located 2.6 meters above ground level. Its inside dimensions are 5.5 meters in length, 2.7 meters in width, and 2.2 meters in height. The tank, with a capacity of 30,000 litres, serves as the primary water source for the spawning, incubation, and storage tanks, ensuring an ample supply. Furthermore, installing a smaller overhead tank holding 5,000 litres is advisable to support these operations.

The spawning pond is constructed using masonry or concrete and has a circular shape. It has an inside diameter of 8 meters and can hold up to 50 cubic meters of water. The depth of the pond at the periphery is 1.20 meters, gradually sloping down to 1.50 meters at the centre. Along the outside of the pond's wall, a water supply line is installed. The entrance to the pond is situated at 14-16 evenly distributed points, positioned at a 45° angle from the tank's radius. This entrance consists of a 20 mm diameter pipe with a nozzle opening, all oriented in one direction. The nozzles are affixed to the vertical wall of the pond, their openings aligned flush with the cement plaster surface near the bottom around the pond's edge. The valve on the inlet pipe allows fertilized eggs and water to be transferred into the incubation pond for hatching. The water flow in the spawning pool creates an artificial riverine condition to facilitate fish breeding. A shower and a perforated galvanized iron pipe are installed to increase the dissolved oxygen levels. The spawning tank can accommodate approximately 70 kg of males and 70 kg of females, yielding around 10 million eggs in a single breeding operation (Szabo *et al.*, 2019).

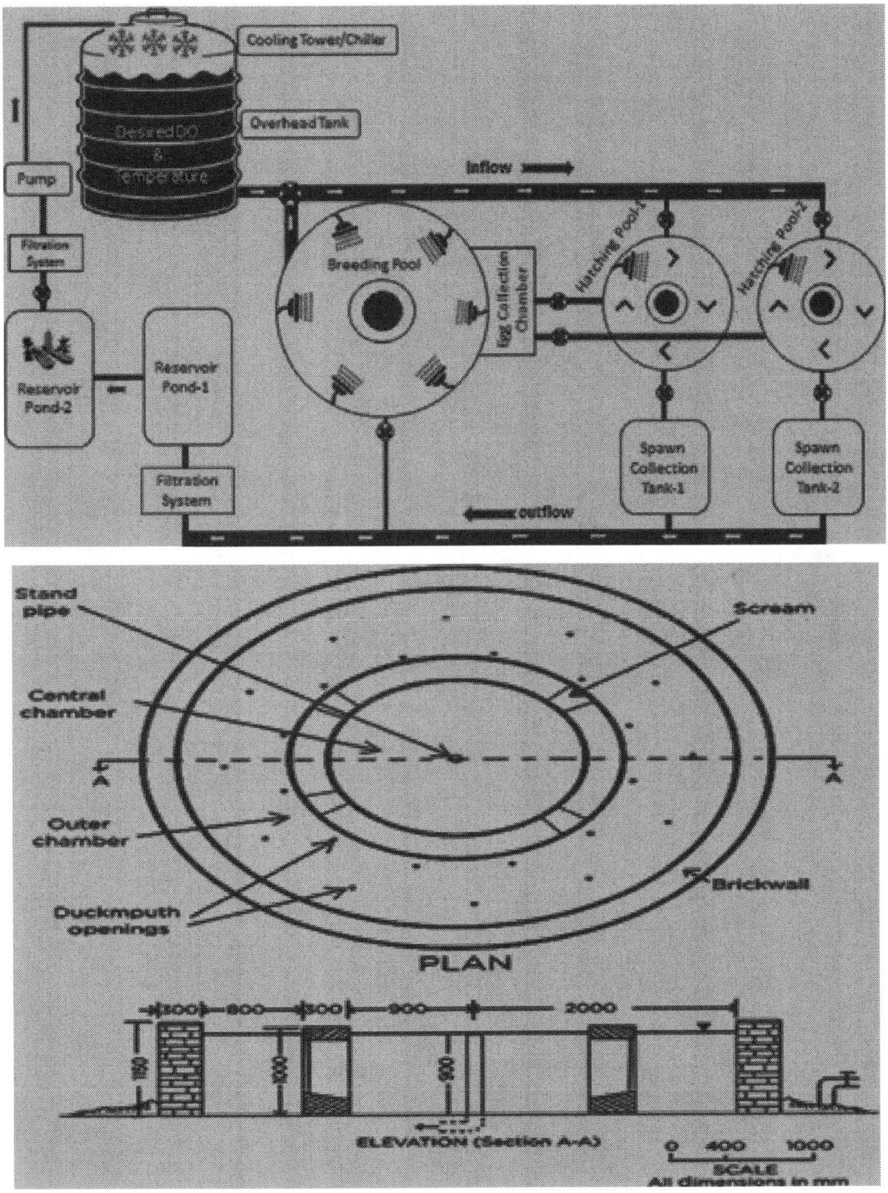

Picture: 1 Chinese hatchery
(Source: https://cl.gy/Zyblm, Rasal *et al.*, 2023)

Incubation ponds consist of two circular ponds with an internal diameter of 3.6 meters each. Each pond is divided into two chambers. The outer chamber is enclosed by a masonry/concrete wall with a dimension of 4 meters in diameter. At 0.76 meters from the outer wall, another circular wall has a fixed nylon screen. These ponds have a depth of approximately one meter and can hold 9-12 cubic meters of water. The setup is structured to handle a vast quantity of 70,000 million eggs within each cubic meter. Inside the pond are strategically placed vertical outlets, each with a diameter of 10 cm. These outlets feature perforations at different levels, facilitating the controlled release of surplus water from the incubation pond. The spawn, accompanied by water, is directed from these incubation ponds to the spawn collection pond through this system (Belonio, 2006).

The facilities of water system consists of various pipelines and ponds. The initial pipeline, starting from the overhead tank, has a diameter of 7.5 cm. The system begins with a more extensive diameter pipeline, eventually reducing to a 5 cm pipe, then further down to 1.2 cm. Within the incubation pond are eight outlets featuring duck mouth openings angled at 45° towards the inner wall, fixed unidirectionally. These outlets receive water from a 10 cm pipeline branching from the circular spawning tank, which divides into two 10 cm pipelines, one for each incubation tank. These pipelines connect to the duck mouth outlets on the pond floor.

To facilitate the passage of hatchlings into the hatchling receiving pond and to drain the outer chamber of the incubation pool fully, there is a 7.5 cm outlet. The targeted water flow velocity in the system ranges between 0.2-0.3 m/sec. The hatchling receiving pond, a rectangular concrete tank, measures 4 x 2.5 x 1.2 m and is positioned at a lower elevation than the incubation pond to allow gravitational drainage if ground levels permit. Freshwater supply from the overhead tank enters through a 7.5 cm pipeline, then splits into three 3 cm pipelines for aeration. Spawn transfer and regulation from the incubation ponds to the spawn receiving pond are managed by 7.5 cm diameter pipes. Additionally, hooks on two opposite walls of the pond support nets for spawn collection, while steps ease access during this process. Excess water from the spawn receiving pond is directed to an open drain and can be utilized in earthen ponds if feasible (Mohale *et al.*, 2020).

The Chinese hatchery operates as follows:

1. Brooders are conditioned in the spawning pond for 4-8 hours.
2. The first injection is given to the females between 4-6 PM.
3. Six hours later, the females receive a second injection, while the males receive a single dose.
4. Water jets commence four hours post-injection to induce circular water motion.
5. Breeding occurs between four to eight hours following the second injection.
6. Each operation treats one crore (10 million) eggs.

7. Eggs are gathered from the bottom and conveyed to incubation pools via pipelines.
8. The water in the incubation pools undergoes further agitation.
9. Within four days, the spawn increases to approximately 6 mm in diameter and is transferred to the hatching spawn receiving pool.
10. The spawn is placed in separate water ponds until it matures into the fry stage.
11. If oxygen is low, a compressor powered by a 1HP motor can provide aeration at a 6 kg/cm^2 rate in the incubation pool (Belton, 2012).
12. Additional aeration methods, such as water showers or jets, can be used as needed.

Indeed, here are the paraphrased versions of the provided sentences:

13. Breeding and hatching activities can occur in approximately 30 sets of 4-day cycles within the 120-day breeding window.
14. One batch can hatch roughly ten million eggs.
15. With a hatching success rate of 95%, approximately 285 million spawns, each about 6 mm in size, can be generated.
16. The hatchery features permanent structures, is easy to operate, and requires a minimal workforce.

3. D- variety hatcheries:

Seed production depends on nature, but the issue has been resolved with the CIFE-D-81 hatchery model. This modern hatchery allows fish breeding without relying on rainfall. Consequently, we are no longer reliant on the monsoons and natural environment. The breeding unit serves as the residence for the brooders, while the hatching process takes place within controlled jars, ensuring regulation of silt levels, oxygen content, temperature, and metabolites. Within the hatchery system, there are both breeding and hatchery units.

The breeding unit comprises air conditioners, breeding tanks, sprayers, a water circulation system, an aeration mechanism, water pumps, overhead tanks, and a filtration setup. These are situated within a climate-controlled environment, maintained by a 1.5-ton air conditioner, covering an area of 22.5 square meters. Two breeding tanks, each measuring 440 x 115 x 80 cm, are utilized for breeding 240 kg females through 30 operations over four months during the breeding season. These tanks can be constructed from plastic pools, LDPE, cement, or fibreglass. Equipped with fine 75 mm diameter showers and spray channels along their upper edges, these tanks allow for independent or simultaneous operation of the spray and showers as required. A 1/16 HP pump facilitates water recirculation within the breeding tanks, ensuring oxygenation through the spray and showers. Additionally, each breeding tank houses two floating hapas measuring 180 x 90 x 90 cm, containing close-net hapas with dimensions of 170 x 80 x 80 cm and a mesh size of 20 mm. These nets feature an opening for introducing injected brooders, enabling the system to yield 2.4 million eggs in a single operation.

Water from reservoirs, ponds, or tube wells is pumped directly through a filtration system to eliminate silt and suspended solids and then directed into overhead tanks. From there, it is distributed to breeding tanks via spray and showers, enhancing dissolved oxygen levels, maintaining water temperature, and mimicking natural environmental conditions. Additionally, aeration is facilitated through an oil-free air compressor or blower (Arfken *et al.*, 2021).

The hatching facility consists of overhead tanks, vertical hatchery jars, an oil-free air compressor and blower, spawneries, spray systems, and floating hapas. Typically housed in a shed or building, the hatchery is carefully regulated to maintain a temperature range of 27°-29°C. Aeration mechanisms ensure dissolved oxygen levels remain between 7-9 ppm. The hatchery jars, constructed from low-density polythene, stand at a height of 62.5 cm with a capacity of 40 litres, featuring a 37 cm diameter pipe with a control valve fitted below each jar and an independent control valve for each. These jars are arranged in series, with an inner egg vessel of 20 litres capacity placed inside to facilitate the removal of egg shells post-hatching. Every set of three jars is accompanied by a spawn-receiving tank made of low-density polyethene, boasting a capacity of 1450 litres, 6′ in diameter and 3′ in height. Water spray systems are positioned around the upper edge of each tank to support the hatching process.

Spawn receiving tanks have Overflow pipes measure 50 mm in diameter and are linked to a storage tank in the tanks where the spawn is received. From there, water is pumped back to the overhead tank via a filter for recirculation. A floating hapa made of nylon with a fine mesh is positioned in the spawn receiving tank to secure the spawn. Transfer of spawn from hatchery jars to the hapa is facilitated using a flexible PVC pipe of 32 mm diameter to ensure the spawn remains intact. Water cooling and aeration are achieved through showers and sprays. Aeration systems are installed in the hatchery jars, and the spawn receives hapa to enhance dissolved oxygen levels. Furthermore, eggs are kept buoyant in the egg vessel (Somerville *et al.*, 2104).

D-81 hatchery units, chosen breeders undergo induced breeding and are introduced into the breeding enclosures. If the water temperature exceeds optimal levels, gradual acclimatization is employed by lowering it to 26°-27°C within the breeding facility. Once the temperature stabilizes at 26°C, the air-conditioning is deactivated while the misting and showers persist.

Upon successful breeding, the large meshed enclosure and the exhausted breeders are dismantled, leaving the eggs within the breeding section of the enclosure. Following a 5-hour interval, the eggs are shifted to the hatching facility. Within 4 hours of spawning, the eggs are relocated to the egg vessel, securely positioned within the hatchery jar. Each hatchery jar can hold approximately 2 to 2.25 lakh eggs, depending on the species.

A consistent, gentle flow of air and water is maintained in the jars to allow the eggs for free motion. The flow rate of water is kept between 1-2 litres per minute. Typically, the eggs hatch within 14 hours. Once hatching is finished, the egg container and the shells are removed. The water flow in the jars is then slightly increased to aid in transferring the hatchlings into the spawn receiving tank. Any remaining hatchlings are transferred into the hapa using a 25 mm diameter pipe for siphoning.

After the jars are emptied, the water flow is halted. A spray system is set up around the top edge of the spawn receiving tank and kept operational to maintain optimal levels of dissolved oxygen and low temperature. The aeration and spray systems continue to operate continuously until the yolk sacs of the hatchlings are absorbed, typically lasting around two days. The hatching success rate normally falls between 93% to 98% (Arfken *et al.*, 2021).

Advantages:

1. This system utilizes low-density polyethene, which makes it highly durable and resistant to breakage.
2. It boasts easy packing and transportation capabilities, allowing for convenient movement to various interior locations.
3. A controlled temperature system has been implemented to ensure optimal conditions.
4. The circulating water undergoes filtration to remove metabolites and maintain water quality effectively.
5. By introducing additional aeration, the oxygen levels in the water are elevated to 7-9 parts per million (ppm).
6. Remarkably, the hatching rate remains high even with low fertilization rates, indicating the system's efficiency.
7. This system guarantees uninterrupted breeding and hatching processes, regardless of the absence of rain or monsoon seasons.
8. The filtration process keeps the water clear of sediment and silt, promoting a pristine and healthful setting.
9. Each jar has independent regulation mechanisms for aeration and water flow. This enables the isolation of any jar facing issues such as mortality, pollution, or disease from the rest of the system.
10. In this system, the typical incubation period for ordinary carp eggs, usually 72 hours, is significantly reduced to 42 hours (Brannon *et al.*, 2004).

4. Earthen pot hatcheries

One of the oldest methods adopted for hatching involves using locally made-earthen pots. The collected eggs are placed inside these pots, and hatching occurs. The temperature and pH levels within the pots experience moderate fluctuations. However, it is essential to note that this method is not widely favoured or popular among hatcheries. The percentage of successful hatching achieved through this method is approximately 40% (Priyadarshi *et al.*, 2017).

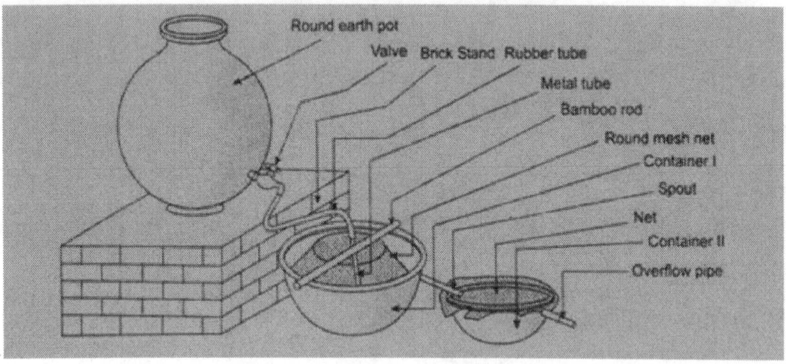

Picture: 2 Earthen pot hatcheries (https://www.notesonzoology.com/india/fishery/4-important-types-of-hatchery-fisheries/731)

5. Cement hatching:

The cement-lined hatching pits serve as containers for incubating eggs, providing advantages such as cost-effectiveness, operational simplicity, and a steady water supply. Nevertheless, they require significant initial investment, and mortality rates are often linked to issues like oxygen depletion and increasing water temperatures. Studies indicate a hatching success rate from 30% to 50% (Bisht *et al.*, 2013).

6. Hatching hapas:

Double-cloth hatching hapas are extensively utilized in aquaculture practices. These hapas are securely positioned in shallow water using bamboo poles. The hapa structure consists of two layers, with the outer layer crafted from thin or coarse muslin cloth. In contrast, the inner layer is constructed from circular mesh mosquito netting fabric. The standard dimensions of the outer wall of the hatching enclosure are around 2 meters in length, 1 meter in width, and 1 meter in height.

In contrast, the inner wall is slightly smaller, measuring approximately 1.5 meters in length, 0.75 meters in width, and 0.9 meters in height. The water depth is regulated at about 30 centimetres to maintain ideal conditions. These enclosures are set up in a series.

Each enclosure holds 75,000 to 100,000 eggs within the inner wall for hatching. After hatching, the newborns navigate through mosquito netting and move into the outer enclosure, leaving eggshells and spoiled and non-viable eggs behind. Following this, the inner enclosure is removed, and the hatchlings remain in the outer enclosure for roughly 40 hours until they have absorbed the yolk sac. The success rate of hatching typically falls between 40% to 50%.

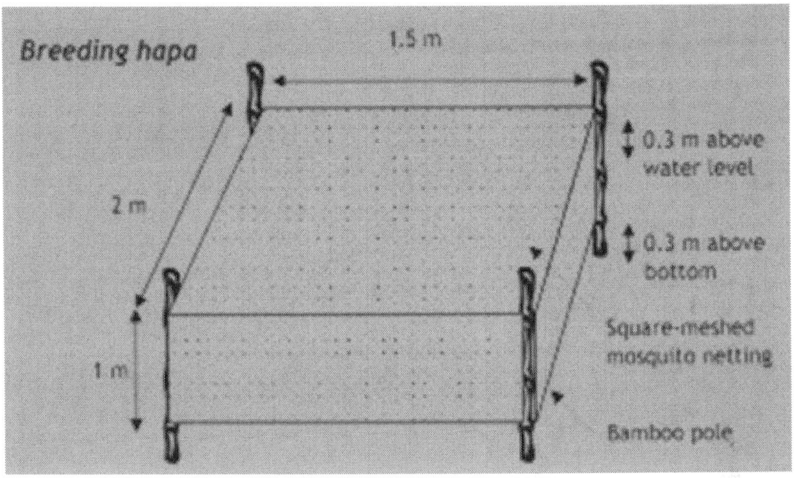

Picture: 3 Hatching hapas (https://images.app.goo.gl/m9vDerzrUssoxpTx5)

Garfield hatching hapas The primary benefits of utilizing Garfield hatching hapas over cloth hapas include significantly reduced costs and the prevention of egg pollution and potential mortality by keeping them away from Earth. However, there are several drawbacks linked with Garfield hatching hapas. Initially, pore blockage due to silt accumulation can lead to high mortality rates. Additionally, the susceptibility of these hapas to damage by crabs further increases mortality. Furthermore, Garfield hapas have a relatively short lifespan of around two years. Weather variations can exacerbate mortality rates, and these hapas necessitate a larger water supply. Notably, the hatching percentage with Garfield hatching hapas is around 50-60% (Devi *et al.*, 2009).

7. Floating hapas

Picture: 4 Floating hapas (https://priyafil.com/aquatech/fish-breeding-nets-hapas)

Floating hapas are an improved alternative to traditional hapas, specifically designed to accommodate fluctuations in water levels. Unlike conventional hapas, floating hapas can be easily installed in rocky areas without bamboo poles. They are also suitable for deeper water sections, allowing a gentle water current to pass through the hapa, promoting improved water exchange and aeration. While similar in concept to conventional hapas, floating hapas are constructed using frames made of polythene or aluminium pipes, with floats attached for buoyancy. The hapa is secured to fixed objects using long ropes to prevent drifting with the current. Additionally, the floating hapa is collapsible and can be easily assembled. The outer hapa measures 2 x 1 x 1 meters, while the inner hapa is slightly smaller at 1.75 x 0.75 x 0.5 meters. However, there are specific challenges associated with floating hapas. Silt deposition within the hapa can lead to spawning mortality, and the process of repositioning and rearranging the hapa can be time -consuming. Moreover, the hatching rate of spawn is not exceptionally high, averaging between 50% and 70% (Altamirano *et al.*, 2021).

8. Tub hatchery

This hatchery, implemented in Madhya Pradesh, represents an advancement from traditional fixed hapas by enabling hatching in flowing water. It operates by employing a continuous water flow facilitated by gravity and siphons. The setup comprises a series of interconnected galvanized iron hatching tubs, typically numbering between 8 and 12, ensuring a consistent water flow. Each tub, measuring 2.5 feet by 2.5 feet by 1.5 feet, is equipped with both outer and inner nets. Fertilized eggs are placed within these tubs for hatching, with the hatching success rate typically falling between 50% and 70%. Given the operational characteristics of this system, constant monitoring is imperative to uphold its functionality (Palupi *et al.*, 2021).

Picture: 5 Tub hatchery
(https://priyafil.com)

9. Cement cisternae hatchery

The traditional tub hatchery has been replaced with a new cement cistern system. Instead of tubs, these cisterns are now utilized, positioned beneath the dams of the dry bundh. Water from the ponds is redirected to these cisterns to support the hatchery operation. Each cistern measures 2.4 x 1.6 x 0.45 meters and is organized in two rows. Unlike the previous tub setup, these cisterns operate independently, featuring separate inlets and outlets. Each cistern accommodates around 300,000 eggs for hatching. The hatching rate ranges from 50% to 70% (Van *et al.*, 1992).

10. Vertical jar hatchery:

By employing this improved technique, a remarkable 90% survival rate of fish hatchlings can be achieved. The depicted hatchery is outfitted with crucial elements such as a dependable water source, a breeding tank, equipment for incubation and hatching, and a spawner. The vertical jars used in this setup are constructed using glass, polythene, and iron. These materials ensure the necessary durability and functionality of the jars within the hatchery system.

- The jar hatchery presents a notable benefit due to its minimal water usage. With a cluster of 40 jars, it can manage 2 million fertilized eggs daily, utilizing only 20,000 litres of water.
- Another standout characteristic is its capacity to function within a confined space. The footprint required for a 40-jar setup is around 10 square meters, at most 20 square meters, ample for hatching 2 million eggs. In contrast, hatching hapas in ponds would require 150 square meters of space.
- During the summer, when water temperatures soar above 32° Celsius, hatching in hapas can be adversely affected. However, in jar hatcheries, this challenge can be overcome by air-conditioning the room.
- The jar hatchery allows for visual inspection of developing embryos, enabling necessary adjustments as per requirements.
- A set of 40 jars, along with accessories, costs Rs. 10,000. These jars have a lifespan of 10 years, resulting in an annual cost of Rs. 1,000 for hatching 2 million eggs. On the other hand, handling the same number of hatchlings in hapas costs Rs. 9,000. Hapas lasts only two years and requires more labour. Hence, the jar hatchery is more convenient and economically feasible (Bisht *et al.*, 2013).
- The jar hatchery can hatch 2 million eggs in one day within roughly 20 square meters, with a survival rate of about 90%. During the monsoon season, the hatchery can manage around 200 million eggs.
- Another advantage of the jar hatchery is that it allows successful breeding in the same air-conditioned room. Breeders thrive in temperatures ranging from 26° to 28° degrees Celsius.

- Adverse water conditions found in the jar hatchery can be corrected. For instance, when reservoirs experience increased hydrogen sulfide content during summer, it affects hatching in hapas supplied with such water from the ponds. One potential solution is to purify the water stored in the overhead tanks before distributing it to the hatchery jars.

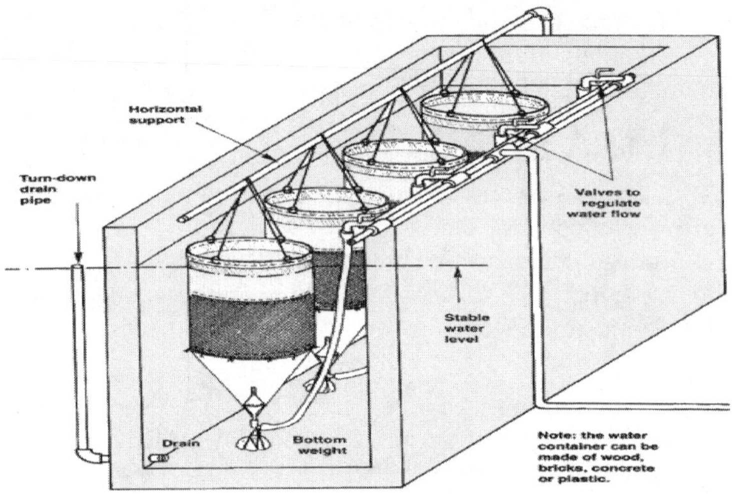

Picture: 6 Vertical jar hatchery
(https://www.fao.org)

Disadvantages of glass:

- Fragile, prone to damage
- Difficult to move or relocate
- Susceptible to breakage during transportation
- Lack of temperature control system
- Metabolites not removed from circulating water
- Insufficient air circulation

The "Transparent polythene sheet hatchery" is a system in which glass jars are substituted with polythene containers. These containers are cylindrical, measuring 27 cm in height and 10 cm in diameter, and have a capacity of 2 litres each (Mohan *et al.*, 2005).

In the giron jar hatchery, a switch is made from glass to galvanized jars, offering several advantages. These galvanized jars are more resilient, cost-effective, and have a larger capacity. Additionally, they are better suited for the specific conditions found in local villages. These containers take on a conical form and are equipped with a small spout positioned at the upper section, functioning as a dispensing point. Each jar stands at a height of 75 cm with a diameter of 23 cm. To secure the jars, they are affixed within an angular iron framework. A consistent rate of approximately 1 litre per minute is maintained regarding water flow.

11. Plastic bin hatchery

This group comprises eight hatchery cum spawner units (HCS units) accompanied by a 5,000-liter water tank. The tank receives water from a natural source via pumping and is linked to the inlet pipelines of each unit, facilitating easy sequential connection. Inside this hatchery are 200 million eggs designated for hatching, with a success rate ranging from 70% to 80%.

Each unit comprises an outer enclosure and an inner shared egg vessel. The exterior hatchery container is a rectangular tub crafted from aluminium sheet, with dimensions of 54" x 18" x 22", and boasts a capacity of 243 litres. It is divided into three chambers with unequal sizes. In each hatchery unit, 8 litres of eggs are placed for hatching at a time. The container also has an inlet, outlet, and drain pipe.

The typical egg vessel used in this system is constructed using a 14 gauge aluminium sheet featuring perforations with a diameter of 2.5mm. Each outer container holds three of these egg vessels. The container is cylindrical, measuring 12 inches in diameter and 12 inches in height. It features a mechanism comprising a plunger and lid designed to adjust to various requirements. This mechanism is mounted on a vertical aluminium rod, equipped with multiple holes spaced at 1 cm intervals, allowing the plunger-lid arrangement to be fixed at different heights as needed. The lid closely covers the eggs within the vessel, preventing overflow and facilitating efficient water circulation. The capacity of each egg vessel is approximately 200,000 eggs. One notable advantage of this system is its cost-effectiveness, primarily due to using plastic materials. Additionally, it is designed to be user-friendly and easy to operate.

However, there are certain disadvantages associated with this system. Firstly, it lacks a temperature control device, which may be crucial for maintaining optimal conditions for egg incubation. Moreover, there is no provision for additional air circulation, which could potentially affect the overall environment within the vessel. Furthermore, the system does not effectively remove metabolites from the circulating water, which can adversely affect the eggs. Finally, there is a risk of rhegaplankton contamination from the overhead tank, and these contaminants can be harmful to the developing spawn.

12. Plastic bucket hatchery

The phage system comprises a sturdy plastic bucket encompassing an aluminium bin egg vessel equipped with perforations and a spawner constructed from galvanized iron sheets. The plastic bucket stands at a height of 47 cm with a diameter of 30 cm, boasting a capacity of 45 litres. The container is equipped with three openings located at the bottom and two exits placed at the top. The eggs are delicately placed within the egg receptacle to aid in hatching. Notably, the survival rate within this phage system ranges from 70% to 80% (Gonzalez *et al.*, 2021).

13. Hatching dipnet hatchery

This hatchery facility comprises a spawning tank, two hatching tanks, two breeding tanks, and an overhead tank. The spawning tank measures 2.36 x 3.23 x 0.9 meters, the hatching tanks are 3.3 x 1 x 1 meter each, and the breeding tanks are 1.2 x 0.7 x 1.06 meters. Water is sourced from the overhead tank positioned at a height of 3.2 meters above the roof. All tanks have inlet and outlet pipes and sprayers for water distribution. Air coolers are utilized for temperature regulation. The hatching tanks employ barrel-shaped nets with steel rings measuring 65 cm at the top and 46 cm at the bottom. These nets are covered with 1/16 inch mesh cloth, featuring a 50 mm brass spray head at the bottom. Each net accommodates approximately 100,000 eggs, with a 1-1.5 litre water flow per minute maintained during hatching. Hatchlings transition to the spawning tanks, achieving an average hatching rate of around 80% (Bernier *et al.*, 2005).

14. Circular cisternae hatchery

A drum made of galvanized iron sheet has a 1m diameter and height. It has an inlet pipe fixed at a 45° angle, 5cm above the bottom, connected to the water supply. A check valve near the inlet regulates the water flow. The inlet pipe initiates water movement within the drum, facilitating circulation. Any extra water is discharged through the upper outlet. Within the drum, eggs are present and are carried around by the flowing water. A mesh-covered outlet ensures that eggs cannot drift away. Following hatching, eggshells dissolve and are carried away with the excess water. The newborns stay within the drum and are gathered at a later time. This water circulation ensures sufficient oxygen levels, leading to a high hatching success rate of 90% (Atta *et al.*, 2012).

Picture: 7 Circular cisternae hatchery
(https://nfdb.gov.in)

References

Altamirano, J. P., Sinsona, M. J., Caasi, O. J. C., Uy, W. H., Noran-Baylon, R. & Juinio-Meñez, M. A. (2021). Factors affecting the spatio-temporal variability in the production of sandfish *Holothuria scabra* juveniles in floating hapa ocean nursery systems. *Aquaculture, 541*, 736743.

Arfken, A., Song, B., Allen Jr, S. K. & Carnegie, R. B. (2021). Comparing larval microbiomes of the eastern oyster (*Crassostrea virginica*) raised in different hatcheries. *Aquaculture, 531*, 735955.

Atta, K. I., Abdel-Karim, A. E. & Elsheikh, E. H. (2012). Ultrastructural study of the effect of heavy metals on the regenerating tail fin of the teleost fish, *Oreochromis niloticus*. *The Journal of Basic & Applied Zoology, 65*(4), 232-239.

Belonio, A. T. (2006). *In: Dictionary of agricultural engineering*. Department of Agricultural Engineering and Environmental Management College of Agriculture Central Philippine University Iloilo City, Philippines.

Belton, B. (2012). Culture, social relations and private sector development in the Thai and Vietnamese fish hatchery sectors. *Asia Pacific Viewpoint, 53*(2), 133–146.

Bernier-Bourgault, I., Guillemette, F., Vallée, C. & Magnan, P. (2005). A new incubator for the assessment of hatching and emergence success and the timing of emergence in salmonids. *North American Journal of Fisheries Management, 25*(1), 16-21.

Bisht, A., Anand, S., Bhadula, S. & Pal, D. K. (2013). Fish seed production and hatchery management: A Review. *New York Science Journal, 6*(4), 42–48.

Brannon, E. L., Amend, D. F., Cronin, M. A., Lannan, J. E., LaPatra, S., McNeil, W. J. & Westers, H. (2004). The controversy about salmon hatcheries. *Fisheries, 29*(9), 12–31.

Das, S. K. (2003). Breeding of carp using a low-cost, small-scale hatchery in Assam, India: A farmer-proven technology. *Aquaculture Asia, 8*(1), 8–10.

Devi, G. A., Devi, G. S., Singh, O. B., Munilkumar, S. & Reddy, A. K. (2009). Induced spawning and hatching of *Osteobrama belangeri* (Valenciennes) using ovatide as an ovulating agent. *Asian Fish. Sci, 22*(4), 1107-1115.

Dickson, M. H. & Fanelli, M. (2013). Geothermal energy: utilization and technology. Taylor and Francis Group. https://doi.org/10.4324/9781315065786

González, K., Warkentin, K. M. & Güell, B. A. (2021). Dehydration-induced mortality and premature hatching in gliding treefrogs with even small reductions in humidity. *Ichthyology & Herpetology, 109*(1), 21-30.

Mohale, H. P., Sarang, N. & Desai, A. Y. (2020). The Common Carp and its Culture System Management.

Mohan, D. (2005). Evolution of fish hatcheries in India. *Changing Faunal Ecology in the Thar Desert*, 173.

Palupi, M., Fitriadi, R., Dadiono, M. S., Yonarta, D. & Kusuma, R. O. (2021). Profitability analysis of tiger grouper hatchery (*Epinephelus fuscoguttatus*) household scale in Kelatakan Village, Situbondo, East Java. *Journal of Aquaculture and Fish Health, 10*(2), 192-198.

Priyadarshi, H. & Singh, A. A., Jamoh, N., & Chakraborty, P., Bogi, R., Singh, S. B., Mallik, A. Amom, M., Khuman, O.N., Paul, T. & Das, R. (2017). Identification of key factors to negotiate spontaneous spawning of *Clarias batrachus* in confinement. Fishery Technology. *54*. 8-13.

Somerville, C., Cohen, M., Pantanella, E., Stankus, A. & Lovatelli, A. (2014). Small-scale aquaponic food production: integrated fish and plant farming. *FAO Fisheries and aquaculture technical paper, 589*.

Szabó, T., Urbányi, B., Müller, T., Szabó, R. & Horváth, L. (2019). Assessment of induced breeding of major Chinese carp at a large-scale hatchery in Hungary. *Aquaculture Reports, 14*, 100193.

Van Winkoop, A., Booms, G. H. R., Dulos, G. J. & Timmermans, L. P. M. (1992). Ultrastructural changes in primordial germ cells during early gonadal development of the common carp. *Cell and tissue research, 267*, 337-346.

CHAPTER 3
SEED PROCUREMENT

Introduction

Fish seed is of utmost importance for the successful practice of fish culture. Our nation has abundant freshwater reservoirs designated for pisciculture & spanning. There are approximately 2.85 million hectares of ponds and tanks. Additionally, an extra 2.05 million hectares of aquatic terrain exists in the guise of reservoirs or lakes. To reach maximum utilisation of the cultivable expanse comprising i.e. 2.85 million hectares, it has been projected that an immense quantity of around 14,250 million fries would be indispensable. This projection is grounded on a prudent stocking rate of 5,000 fry per hectare. The fish fry production currently stands at 15,007 million, surpassing the anticipated requirement. Nonetheless, it is imperative to acknowledge that this production encompasses the fry necessary for the currently available resources.

Furthermore, an additional quantity of at least 4,100 million fry is imperative to stock the expanses of lakes and reservoirs adequately. This requirement is calculated based on an average stocking rate of 2,000 fries per hectare. These figures indicate the pressing need to augment fry production to sufficiently stock our water resources for fish culture (Erickson & Halford, 2020).

The acquisition of fish seeds is primarily done through three sources: rivers, hatcheries, and bundhs. While extracting seeds from riverine sources has been a traditional method, it is a labour-intensive process that yields a mix of desired and undesired fish seeds. On the other hand, hatcheries offer the most efficient means of obtaining fish seeds. Additionally, bundh breeding provides an effective method for collecting fish seeds by establishing a natural habitat.

The fish fauna in India's river systems exhibit variations in distribution and abundance. The variances between these rivers stem primarily from their distinct ecological settings, encompassing factors like gradient, terrain, flow, depth, temperature, and substrate. Northern rivers boast perennial flows, fostering ideal environments for many fish species to flourish. Consequently, these waterways support robust commercial fisheries. Conversely, except in deltaic zones, the fisheries in peninsular rivers tend to be lacking, spanning both upper and middle reaches (Dedefo *et al.*, 2017).

Natural seed Resources

Major River System

The fish fauna in India's various river systems exhibit variations in their distribution and abundance. The variations in characteristics predominantly stem from the unique ecological settings of each river, encompassing elements such as incline, landscape, current, profundity, temperature, and substrate. The perennial rivers in the north sustain vibrant commercial fishing industries, barring the deltaic zones. Conversely, the peninsular rivers, across their upper and middle stretches, exhibit restricted fishery potential (Badstue *et al.*, 2006)

India boasts five major river systems: the Ganga, Brahmaputra, Indus, East Coast, and West Coast.

The Ganga river system:

The Ganges, coursing through the regions of Haryana, Delhi, Uttar Pradesh, Madhya Pradesh, Bihar, and West Bengal, stands as an immense hydrological network spanning an impressive 8,047 km. It is the largest river in India, home to a diverse and abundant freshwater fish population. The collection of fish eggs takes place both from the breeding grounds and downstream areas. To collect the eggs, individuals disturb the riverbed and scoop them out using a traditional cloth known as a gamcha. The eggs are typically found in water around 1-2 feet deep. The act of gathering spawn on a large scale for commercial purposes is primarily observed in the states above, pivotal in bolstering the nation's overall fish yield by a notable 51.9%.

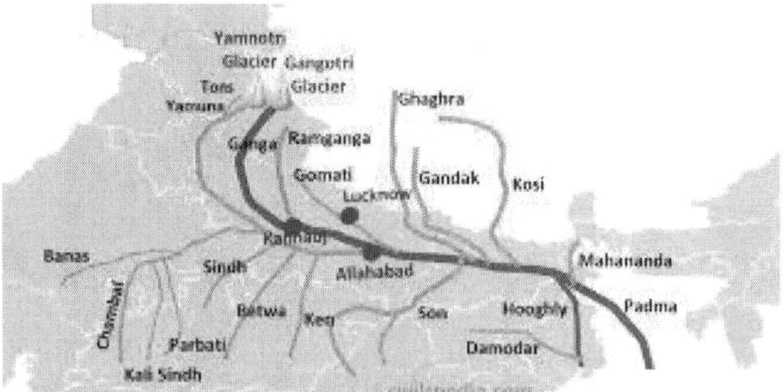

Picture 1: The Ganga river with its tributaries (https://civilspedia.com/)

During the period spanning from May to September, the optimal window for gathering significant carp spawn emerges. This timeframe coincides with the onset of melting snow, especially pronounced in higher elevations, which triggers floods and facilitates the downstream transport of carp spawn. In the Indian context, the initial appearance of spawn is typically noted in the Kosi River, succeeded by the primary Ganga River, Gomati River, and various western tributaries. Notably, billions of carp fry and fingerlings are harvested between July and October in North

Bihar, underscoring the prolific abundance of fish within the River Ganga and its adjacent areas (Bhaumik, 2017).

The Brahmaputra river system:

The river system delineated in the passage meanders through Assam, Nagaland, and Tripura states. Renowned for its swift currents, it provides a sanctuary for crucially valuable major carp species. This river network has diverse characteristics, spanning a vast distance of 4,023 kilometres. Along the northern bank of the Brahmaputra, its tributaries exhibit considerable size and boast rugged, shallow-braided channels with coarse sandy bottoms. Laden with substantial silt deposits, these tributaries contribute to the river's sedimentary composition. Conversely, the tributaries flanking the southern bank are notably deeper, presenting a striking contrast in their topographical features.

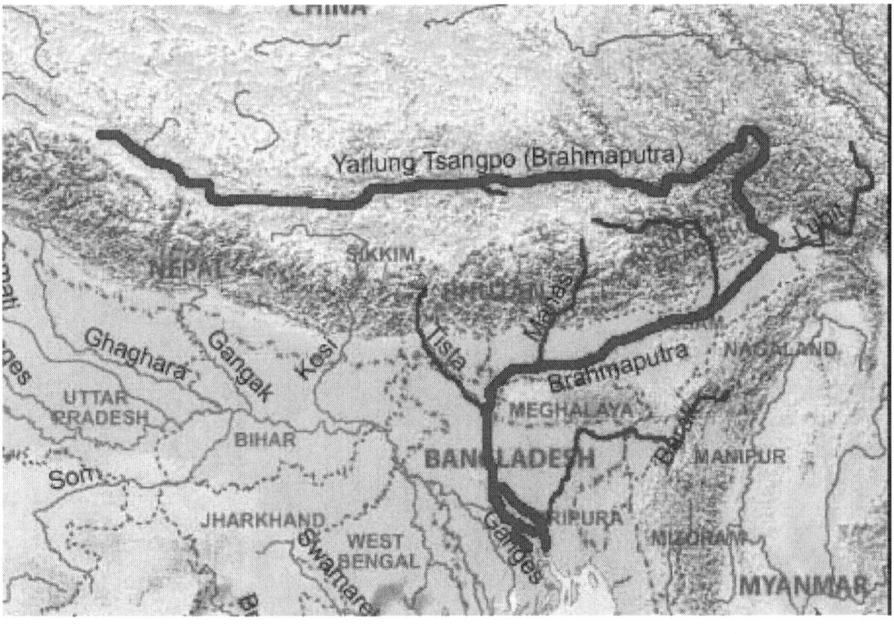

Picture 2: The Brahamputra river and its tributaries (https://lotusarise.com)

In this swift river with steep banks, seeds for major carp are gathered. The procedure entails anchoring two extended bamboo poles close to the banks and tethering a boat against the current. Nonetheless, the efficacy of this method in acquiring major carp remains low. A research endeavour undertaken at the Gauhati Centre in 1969 unveiled that rohu, a particularly significant carp variety, constituted merely 9.58% of the overall catch.

The river's wild and erratic behaviour, driven by the sharp slopes of its feeder streams, swiftly alters its flow. This rapid flux poses a formidable challenge in gathering carp seeds, resulting in a scarcity of viable ones (Sharma, 2022).

Indus river system:

The Indus River system, spanning a vast length of 6,471 km, is a testament to its expansive nature. The Indus River system exhibits considerable opulence when juxtaposed with the Brahmaputra River. Within the territories of Himachal Pradesh, Punjab, and Haryana, the Beas and Sutlej rivers, accompanied by their myriad tributaries, contribute significantly to the landscape. Nevertheless, Himachal Pradesh lacks a notable commercial fishery for major carp due to the prevalence of cold-water environments in the upper reaches of its rivers. Conversely, Punjab emerges as a conducive environment for carp fishery (Shaw, 2005).

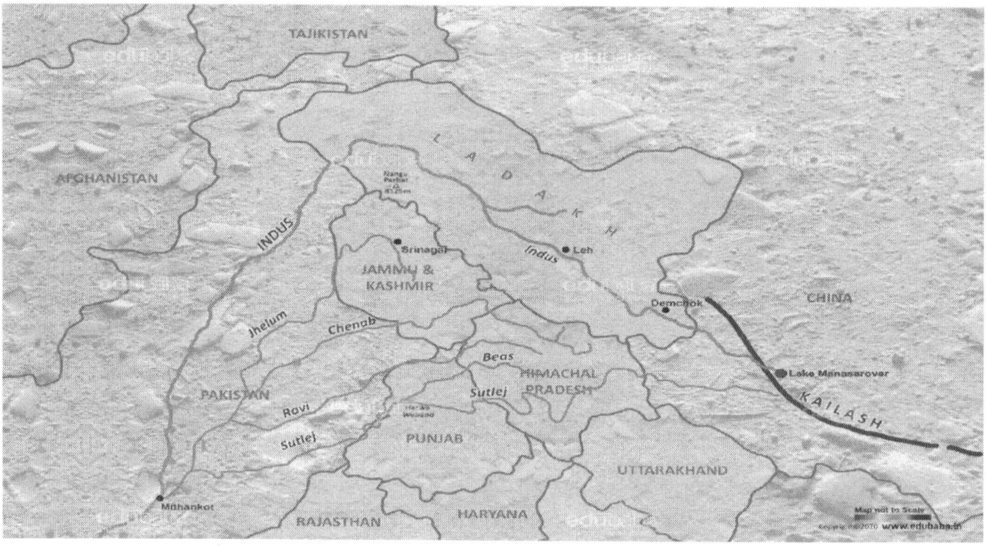

Picture: 3 Indus River System: https://edubaba.in)

East coast river system :

The rivers along India's eastern coastline, such as the Mahanadi, Godavari, Krishna, and Cauvery, course eastward, converging into the Bay of Bengal. Spanning a combined distance of 6,437 kilometres, these rivers form a significant hydrological network. Of these, the Mahanadi stands as Orissa's foremost waterway, crucial for the state's pisciculture, especially as a critical reservoir of fish seed. En route from its inception to Sambalpur, the Mahanadi hosts diverse species of hillstream fishes, enriching the river's ecological tapestry.

Multiple centres for collecting spawn have been pinpointed in the expanse from Sambalpur to Cuttack. Conversely, within Maharashtra, the Godavari river network and the Krishna River form the predominant segment of the eastern coastline river system. This is chiefly concentrated in Maharashtra and Andhra Pradesh. Unlike the situation along the Godavari River in Maharashtra, there are no established spawn collection centres in that area (Smale & Jayne, 2003).

The delta regions of these rivers are known to be highly abundant in fish. "In Rajahmundry, within the Godavari River, major carp spawning accounts for merely 20.3%. The upper sections of the Cauvery River, with their rapid currents and cool climate, are unsuitable for carp fishing. Conversely, the middle and lower stretches of the Cauvery River offer favourable conditions for a significant carp fishery.

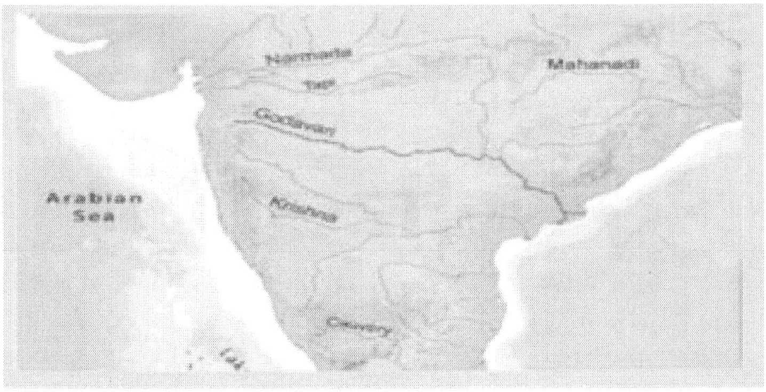

Picture: 4 East coast river system : https://www.slideshare.net

West coast river system:

The principal waterways on the western shores, specifically the Narmada and Tapati rivers, meander through Madhya Pradesh, Maharashtra, and Gujarat states. Their combined length spans 3,380 kilometres. The initial stretches of these rivers traverse rugged, barren landscapes, rendering them inhospitable for seed harvesting. Nevertheless, the downstream sections provide favourable conditions for seed collection.

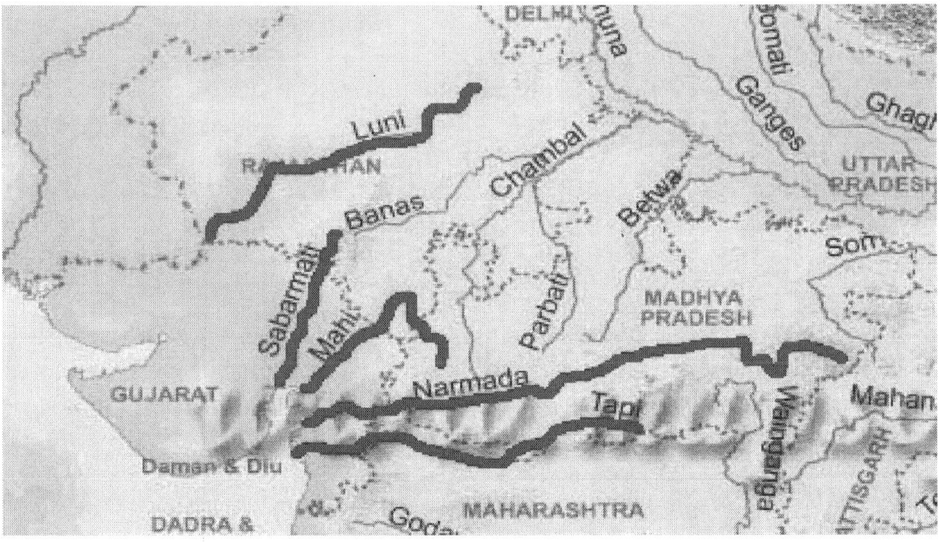

Picture: 5 West coast river system

In India, notable estuarine systems encompass the Hoogly-Matlah estuary of the Ganga River, the Mahanadi estuary in Orissa, the Godavari-Krishna estuary in Andhra Pradesh, the Cauvery estuary in Tamil Nadu, and the Narmada and Tapati estuaries in Gujarat. The nation also boasts significant brackish water lakes such as Chilka in Orissa, Pulicat in Tamil Nadu, and Vembanad in Kerala. A recurring feature across these estuaries is the formation of horseshoe-shaped sandbars at the river mouths. During the southwest monsoon, estuaries receive fresh water from July to October. These estuaries play pivotal roles as reservoirs of freshwater and brackish water fish and prawns (Infantes & Moksnes, 2018).

Lakes and reservoirs:

India boasts an extensive network of naturally occurring lakes and human-crafted reservoirs, offering considerable promise as fisheries. Spanning approximately 2.05 million hectares across the nation, these water bodies hold immense potential. Noteworthy among India's lakes are Chilka, Pulicat, Ooty, Kodaikanal, Nainital, and Logtak, among others. These reservoirs and lakes serve as vital ecosystems and provide crucial habitats for diverse aquatic life forms.

Additionally, India is home to several significant reservoirs. These artificial water bodies are crucial in augmenting the country's fishery resources. Some of the noteworthy large reservoirs in India include Nagarjunasagar, Nizamsagar, Gandhi Sagar, Shivaji Sagar, Tungabhadra, Krishnaraja Sagar, Hirakud, Beas, Govind Sagar, Ramapratap Sagar, Bhavani Sagar, Matatila, Rihand, and Kangasabati, among others. These reservoirs contribute significantly to the fishery sector, supporting the growth of fish populations and providing livelihood opportunities for local communities (Bisht *et al.*, 2013).

Collection of seed from natural resources:

To foster the growth of fish farming in India: It is imperative to ensure the availability of fish seed in substantial quantities. The primary species favoured for cultivation in freshwater ponds and tanks throughout the country are the Indian major carps: Catla (*Labeo catla*), Rohu (Labeo rohita) and Mrigal (*Cirrhina mrigala*). These carp are naturally found in rivers, with their original spawning sites flooded rivers. Throughout history, traditional methods have been established for collecting carp spawn and fry from these natural habitats, primarily originating in Bengal and spreading to other states in eastern India. Even today, specific locations rely on these traditional resources for the fish seed trade (Hasan & Bart, 2006).

To establish a solid scientific foundation, research efforts were launched to explore the potential of seeds in different river systems across India. The objective was to gather valuable information and data to support this endeavour. The investigations aimed to achieve several key objectives. Such as collection nets used for gathering spawn, developing efficient methods for seed collection, and identifying the factors contributing to the fluctuations in fish seed availability in different locations over time. These initiatives enhanced our understanding of seed prospecting and facilitated the sustainable management of fish seed resources in India (Shah, 2021).

Site selection of seed collection:

A pre-monsoon survey is carried out to determine the topography and bank features of the terrain surrounding a specific site. This survey helps assess the extent of the operational area. The investigation also scrutinises the landscape of arid river channels and shoreline attributes to grasp the flow patterns of the river across various flood levels. Moreover, it delves into the arrangement and makeup of fish populations within the designated section of the river, encompassing both indigenous and migratory species. Such data holds significant importance in assessing the prevalence of crucial fish varieties during the monsoon period. Additionally, the study pinpoints the positions of smaller streams, brooks, and artificial waterways, along with their links to the primary river course.

These connecting links play a significant role in the river's breeding grounds. The survey also determines the site's accessibility and identity (Kaiser *et al.*, 1998).

During the survey, it was observed that bends and curves in the river course often exhibit distinct characteristics. One riverside features a precipitous, rapidly eroding bank called the erosion zone. On the contrary, a level, gradually inclining shoreline known as the shadow zone exists. Regrettably, these shorelines are unsuitable for gathering spawn.

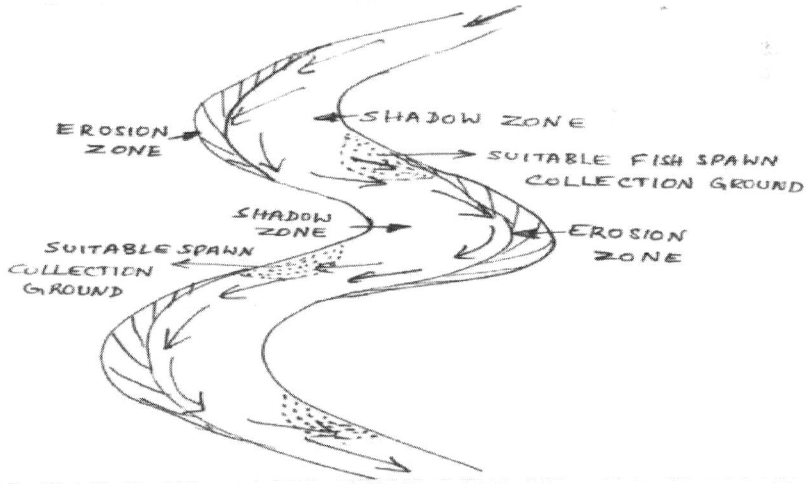

Picture: 6 A river course showing suitable fish spawn collection sites (Source: https://rb.gy/bfaaog)

The prime locations for acquiring fish spawn are situated on the inclined shore. Nevertheless, due to the current's strength, the spawn tends to be displaced towards the peripheries due to centrifugal forces in these regions. Consequently, these zones are well-suited for deploying nets to gather substantial quantities of spawn (Ross *et al.*, 2013).

Methods of seed collection:

Generally, nets designed for seed collection in rivers typically take the form of shooting nets. These nets are constructed in a funnel shape, crafted from finely woven netting material, and strategically positioned with the mouth of the net directed against the current flow. Typically utilised in the shallow peripheries of flooded rivers, shooting nets are equipped with a distinctive feature at their tail end: a stitched-in ring made of split bamboo or cane. Attached to this ring during operation is a receptacle known as a gamcha, a rectangular piece of open cloth. As the marginal current carries the seed particles, they accumulate within the gamcha. Subsequently, these collected seeds are transferred and stored in hapas or containers after extraction from the gamcha.

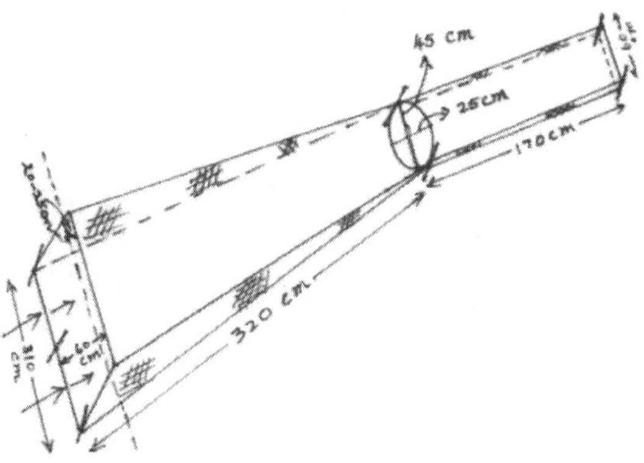

Picture: 7 A typical shooting net used to collect riverine fish spawn. (Source: https://rb.gy/bfaaog)

Picture: 8 A battery of shooting nets ready to be commissioned for riverine fish seed collection. (Source: https://rb.gy/bfaaog)

Benchi jal is a traditional method in Bengal for gathering seeds. It uses a tool called "Benchi jal" to collect seeds. Another tool, the "Midnapur net," is used in Bengal, mainly in the southwest, to gather seeds. The shooting net is fixed in line with the water current. Bamboo poles mark the spot, and the net is tied to them. Two poles are near the entrance, two at the tail ring. The tail ring is wrapped with a "gamcha" at the front end and securely tied. Two more poles keep the gamcha in place (Johannes & Ogburn, 1999).

Multiple trial nets are deployed simultaneously at different suitable locations to determine the most favourable site for spawning availability along a specific stretch of the river. Once pinpointing the optimal spot, the operation proceeds with a full array of nets. After procuring seeds, the contents from the rear end of each net are sequentially scooped every 15 minutes or by the intensity of the spawn. These contents and the gamcha are promptly transferred into a container filled halfway with river water. Following this, the collection undergoes filtration through a mosquito netting sieve to remove undesired organisms and non-floating debris.

The spawn is then measured, placed in hapas for conditioning, and transported to fish farms stocked in nurseries (Meshram *et al.*, 2009).

Picture: 9 Collection of riverine fish seed using a gamcha. ((Source: https://rb.gy/bfaaog)

Picture: 10 Fish seed being collected along the bank of a river (Source: https://rb.gy/bfaaog)

Essential factors affecting seed collection

Flood:

Floods and spawns exhibit a positive correlation, as observed in their relationship. It is common for a season to witness three or more floods. The typical flood pattern involves an initial rise in water levels, followed by a subsequent recession. This cycle repeats, with each flood occurring after a few days. During these floods, carp breed within the rivers (Pilgrim *et al.*, 2007).

The first flood of the season often presents the opportunity to find the spawn of undesirable species. However, it is in the subsequent floods that the central carp seed becomes more abundant. In the periods between floods, the availability of significant carp seed decreases, resulting in fewer catches. Therefore, the presence of spawn is closely tied to the occurrence of floods. As the floods recede, the spawn is carried away from the breeding grounds and downstream in the river. This draining effect happens during the receding phase of the floods. It is worth noting that spawn can be found both during the day and at night, although a greater quantity of seed is typically discovered during nighttime catches (Gomes *et al.*, 2012)

Water current

When the water flow remains gentle at 0.086 km/hr, it exerts no discernible impact on the spawn. Even as the water speed escalates to 0.4 km/hr, the spawn has no notable consequence. Nevertheless, the spawn is swiftly swept downstream once the water velocity surges further. The ideal current velocity for collecting spawn is slow and gentle, ranging from 0.5 to 3 km/hr. In contrast, faster currents in the mid-stream only carry a small amount of spawn, while velocities lower than 1 km/hr are unfavourable for catching spawn. It is essential to mention that spawn availability in the deeper sections of the river is hindered by the lack of floods, which are necessary for spawn production (Rath, 2018).

Distinguishing features of the seed of major Indigenous carps

	Fish Seed	*Catla*	Rohu	Mrigala
1	Eggs	Non -floating Non-adhesive	Non -floating Non-adhesive	Non -floating Non-adhesive
	Diameter	5.3-6.6	5.0	5.5
	Shape	Round	Round	Round
	Colour	Yolk light red	Reddish	Golden
2	Hatchlings	4.67mm	3.7	4.68
	Size	Average	Average	Average
	Yolk sac	Both the bulbous and narrower parts of the yolk sac are equal in length	Like *Catla*	Bulbous and narrower part
	Somites	About 26 pre-anal and 14 post-anal myotomes	Like Catla	28 pre-anal and 14 post and Somites
3	Fry	Dorsal fin rays of more than 11 Head large. No spot at caudal peduncle.	Like *Catla* Head small Transfer band at the caudal peduncle.	Like *Catla* Head small , body slender. A triangular dark spot on the caudal peduncle.
		No barbel, Lips thick, Unfringed	Max. Barbel was present, lips fringed.	The posterior edge exhibits a concave contour in the absence of barbel, with thick lips and an unfringed appearance.
4.	Fingerlings	"Head size is substantial. No markings are present on the caudal peduncle. Absence of barbels."	The head is medium-sized, with a distinct dark band running horizontally towards the tail. It possesses two barbels, one near the maxilla (upper jaw) and the other near the rostrum (snout).	Head moderate, Dark. Spot becomes diamond-shaped barbels, which are not visible.
		Lips are thick and unfringed.	Lips thin and fringed	Lips are thin but not continuous at the corners of the mouth.
		The dorsal, anal, and caudal fins exhibit a rich, dark grey hue.	The dorsal, anal, and caudal fins display a reddish hue accented by murky grey edges.	"The reddish hue adorns the tip of the caudal fin's lower lobe."

Other factors:

The availability of spawn in rivers is not affected by turbidity, pH, or dissolved oxygen. However, turbidity is linked to floods and plays a role in the efficiency of spawn collection. Excessive cloudiness in the water decreases the net's mesh size, necessitating frequent cleaning of the nets. Spawn availability remains unaffected by fluctuations in air and water temperatures. The ideal temperature bracket for spawn collection ranges from 28° to 31° Celsius. Spawn collection is most favourable during overcast conditions accompanied by a breeze and with or without the drizzle."Stormy weather is terrible - it messes up currents and waves and might uproot nets. However, light does not affect spawn collection. "The presence of plankton does not have a direct connection to the availability or abundance of spawn in rivers. Spawn associations are abundant at the beginning of the monsoon season but diminish gradually until they are almost non-existent by the end of the season (Alabaster & Lloyd, 2013).

References

Alabaster, J. S. & Lloyd, R. S. (2013). *Water quality criteria for freshwater fish, 3117,* Elsevier.

Badstue, L. B., Bellon, M. R., Berthaud, J., Juárez, X., Rosas, I. M., Solano, A. M. & Ramírez, A. (2006). Examining the role of collective action in an informal seed system: a case study from the Central Valleys of Oaxaca, Mexico. *Human Ecology, 34,* 249-273.

Bhaumik, U. (2017). Fisheries of Indian Shad (*Tenualosa ilisha*) in the Hooghly–Bhagirathi stretch of the Ganga River system. *Aquatic Ecosystem Health & Management,* 20 (1-2),130–139.

Bisht, A., Anand, S., Bhadula, S. & Pal, D. K. (2013). Fish seed production and hatchery management: A Review. *New York Science Journal,* 6(4), 42–48.

Dedefo, K., Derero, A., Tesfaye, Y. & Muriuki, J. (2017). Tree nursery and seed procurement characteristics influence seedling quality in Oromia, Ethiopia. *Forests, Trees and Livelihoods, 26*(2), 96-110.

Erickson, V. J. & Halford, A. (2020). Seed planning, sourcing, and procurement. *Restoration Ecology, 28,* S219-S227.

Gomes, L. C., Bulla, C. K., Agostinho, A. A., Vasconcelos, L. P. & Miranda, L. E. (2012). Fish assemblage dynamics in a Neotropical floodplain relative to aquatic macrophytes and the homogenising effect of a flood pulse. *Hydrobiologia, 685,* 97-107.

Hasan, M. & Bart, A. N. (2006). Carp seed traders in Bangladesh: Sources of livelihood and vulnerability resulting from fish seed mortality. *Asia-Pacific Journal of Rural Development, 16*(2), 97-120.

Infantes, E. and Moksnes, P. O. (2018). Eelgrass seed harvesting: Flowering shoots development and restoration on the Swedish west coast. *Aquatic botany, 144,* 9-19.

Johannes, R. E. & Ogburn, N. J. (1999). Collecting grouper seed for aquaculture in the Philippines. *SPC Live Reef Fish Information Bulletin, 6*, 35-48.

Kaiser, M. J., Laing, I., Utting, S. D. & Burnell, G. M. (1998). Environmental impacts of bivalve mariculture. *Journal of Shellfish Research, 17*(1), 59–66.

Meshram, S. J., Shingare, P. E. & Ingle, S. T. (2009). Studies on traditional methods of wild giant freshwater prawn seed collection and their potential impact on the aquatic ecosystem. *Asian Fisheries Science, 22*(2009), 185–189.

Pilgrim, S., Smith, D. & Pretty, J. (2007). A cross regional assessment of the factors affecting eco-literacy: implications for policy and practice. *Ecological Applications, 17*(6), 1742-1751.

Rath, R. K. (2018). *Freshwater aquaculture*. Scientific publishers.

Ross, L. G., Telfer, T. C., Falconer, L., Soto, D., Aguilar-Manjarrez, J., Asmah, R. and Zhu, C. (2013). They were carrying capacities and site selection within the ecosystem approach to aquaculture—site *selection and carrying capacities for inland and coastal aquaculture, 19*.

Shah, M. I. (2021). Transportation of Live Fish Seed. *Fisheries and Aquaculture Journal, 12*(6), 1d-1d.

Sharma, N. (2022). Commodities trade, river transport and colonialism: The Brahmaputra river valley in the nineteenth century. *The Indian Economic & Social History Review, 59*(1), 75–94.

Shaw, N. L. (2005). Increasing Native Forb Seed Supplies for. *National Proceedings: Forest and Conservation Nursery Associations, 2004*, 94.

Smale, M. & Jayne, T. S. (2003). *Maize in Eastern and Southern Africa: seeds of success in retrospect* (No. 581-2016-39425).

CHAPTER 4
SEED PRODUCTION TECHNOLOGIES IN FISHES

Fish seed plays a crucial role in fish culture, serving as the primary component for successful cultivation. In our nation, around 2.85 million hectares of ponds and tanks are allocated for fish farming, alongside an extra 2.05 million hectares of reservoirs or lakes. To sufficiently populate the existing 2.85 million hectares with a cautious stocking rate of 5000 fry per hectare, an estimated 14,250 million fry would be needed. Nonetheless, the current fry production exceeds this requirement, at 15,007 million fry. In addition to the existing resources, there is a need for an extra quantity of 4100 million fries to stock the available lakes and reservoirs. The average stocking rate for these water bodies is around 2000 fry per hectare (Soliman, 2017).

Fish seed is acquired from rivers, hatcheries, and bundhs. Gathering seeds from rivers is an age-old yet strenuous technique, yielding a combination of wanted and unwanted seeds. Hatcheries provide the best quality fish seed. Additionally, bundh Breeding is suitable for obtaining seeds by creating a natural habitat.

India's rivers exhibit diverse patterns in the presence and quantity of fish species, influenced by ecological elements like slope, landscape, water movement, depth, temperature, and bed composition. Rivers in the north maintain a constant flow throughout the year, fostering prosperous fishing industries. In contrast, peninsular rivers, excluding their deltaic sections, generally lack significant fisheries in their upper and middle stretches (Agarwal, 2018).

Bundh breeding

In different nations, the preference for pond-reared species in fish farming arises from the ease of collecting and transporting juvenile fish. Conversely, commonly farmed carp species, prized for their fast growth and specific environmental needs, typically do not reproduce within ponds. As a result, their young must primarily be sourced from flooded rivers during the short monsoon season. Indian significant carp usually spawn in flooded rivers between June and August, coinciding with the southwest monsoon period. They may also breed in reservoirs, tanks, and irrigation dams. While mature carp in ponds attain sexual maturity, they do not reproduce. However, transferring these mature breeders from confined waters to semi-confined rain-fed ponds with muddy bottoms prompts Breeding during periods of substantial rainfall and decreasing water temperature (Chattopadhyay *et al.*, 2013).

This observation suggests that ponds may lack the specific conditions conducive to Breeding. Semi-confined rain-fed seasonal water bodies offer higher dissolved oxygen levels, increased light exposure, waves, water flow, turbidity, and lower temperatures, all stimulating fish ovulation. Based on these factors, locations that create riverine conditions with excess rainwater, which stimulate fish ovulation, are called bundhs. These bundhs serve as suitable environments for fish seed production.

Types of bundh

1. **Wet bundh**
2. **Dry bundh**

1. Wet bundh

These wet bundhs, called perennial bundhs, are situated on sloping terrain within a vast catchment area. They include sturdy embankments constructed with an inlet directed towards the higher ground and an outlet at the lower end on the opposite side. Throughout the summer, water is retained solely in the deeper section of the pond, which serves as a breeding ground. The remaining area remains dry and is utilized for agricultural purposes.

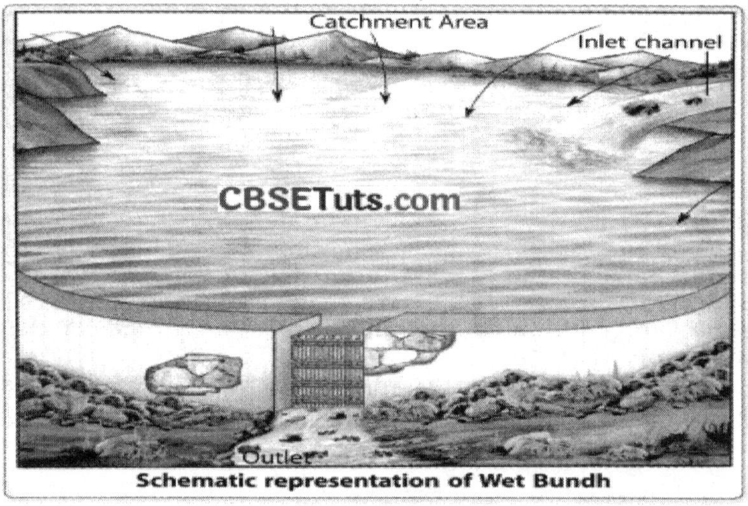

Schematic representation of Wet Bundh

Picture 1: Wet bundh
(https://www.cbsetuts.com)

After intense rainfall, a considerable section of the bundh is inundated as water rushes in from the surrounding catchment area, eventually draining out through the outlet. In this stimulated natural environment, fish begin their spawning activities in the shallow regions of the bundh. To prevent breeders from escaping, the outlet is safeguarded with fencing.

Wet bundhs are considerably larger than their dry counterparts. They share the same designation as perennial bundhs, highlighting their year-round existence and vital role in the local ecosystem (Routray *et al.*, 2007).

2. Dry bundh

A dry bundh is essentially a shallow hollow encircled by an earthen embankment, typically referred to locally as a bundh. It is enclosed on three sides, with the fourth side opening up to a wide catchment area. In the rainy season, bundhs get inundated, yet they stay entirely dry for a considerable duration at other times. These bodies of water are categorized as seasonal rainfed entities and are commonly denoted as seasonal bundhs. The topography of the land influences the location and distribution of dry bundhs. Undulated land is preferred because it offers a larger catchment area, allowing the bundh to fill quickly even with minimal rainfall. Additionally, undulated land facilitates efficient drainage through gravity. In West Bengal, it is deemed optimal to have a catchment area over five times larger than the bundh area, whereas in Madhya Pradesh, a ratio of 1:2.5 is seen as necessary. In the Bankura district of West Bengal, water for most dry bundhs is sourced from storage tanks built in the upland regions.

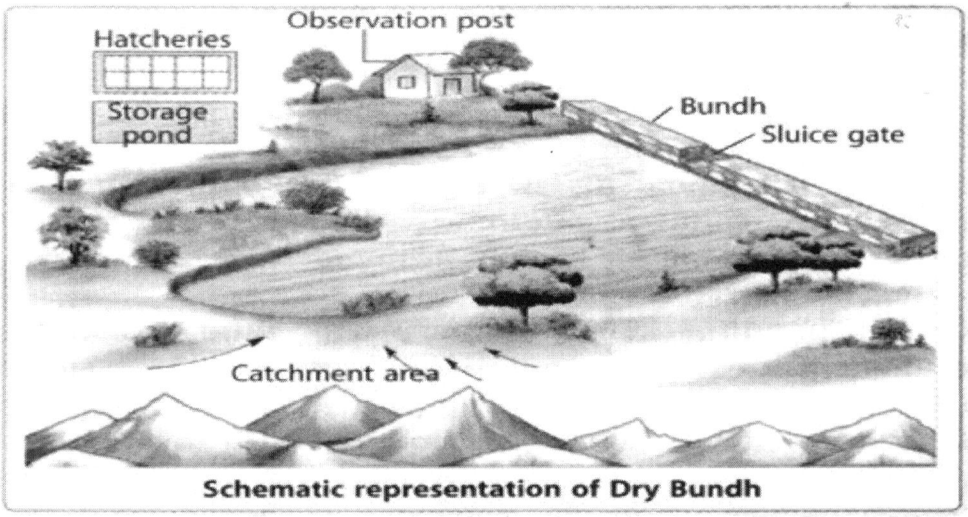

Picture 2: Dry bundh
(https://www.cbsetuts.com)

The practice of bundh Breeding has gained significant importance over the past century. It has gained even more recognition in the last three decades, especially after a review conducted in Madhya Pradesh. In regions such as West Bengal, Rajasthan, and Andhra Pradesh, bundh breeding and river breeding are highly valued and play a significant role in spawn production. Dry bundhs, in particular, are known for yielding 100% pure spawn and are preferred for their simplicity and mass production capabilities (Routray *et al.*, 2007).

Site selection

- An extensive upland area is a source of significant rainwater and sediment flow into the main pond during periods of heavy rainfall.
- The pond ought to have ample shallow marginal sections, serving as ideal breeding grounds for fish.
- The soil should exhibit a gritty consistency, deemed most conducive for fish reproduction.
- The large, shallow expanse of the pond contributes to heightened oxygen levels within the water.

To build a substantial pond with a modest dam, choosing a parcel of land featuring a level surface bordered by steep inclines on three sides is advisable. The fourth side, where water will naturally flow out, should be kept narrow. The inclines on the sides should taper gradually to minimize the amount of construction required or to align with the dam's axis (Gangadhar *et al.*, 2016).

Catchment area

Ensuring a water catchment area of at least 15 hectares of firm ground for every hectare of water surface in the pond is crucial. For areas with soil that retains water well, a ratio of 40 hectares of watershed per hectare of surface water is preferable. It is essential to prevent erosion in fields. If the watershed is either too large or too small, diversion terraces can be employed to rectify the situation. Excess watersheds can be cut off if there is more water, or water can be disposed of elsewhere. A diversion terrace increases the effective watershed for more water.

Collection and handling of eggs

Once the breeding process initiates, immediate arrangements are made for the retrieval and incubation of eggs. To collect the eggs, various materials such as nylon nets, mosquito netting, cloth or gamcha are utilized after lowering the water level. These eggs are then incubated in double-walled hatching hapas, typically positioned within the bundhs. Due to the expansive surface area of wet bundhs, it becomes challenging to retrieve all eggs, resulting in approximately 70% of the eggs being collected. In Madhya Pradesh, egg incubation occurs either in the double-walled hatching hapas installed within the bundh itself or rectangular cement hatcheries measuring 2.4 x 1.2 x 0.3 m.

In contrast, West Bengal employs a distinct method where the eggs are placed in specially dug small earthen pits with mud-plastered walls for incubation. After 12 hours of incubation, the hatchlings are delicately removed from the pits using muslin cloth and then transferred to more giant earthen pits prepared similarly. The survival rate in the hapas is around 35-40%, but by utilizing modern hatcheries, this rate can be significantly increased to 97% (Bagchi & Jha, 2011).

Improved features of dry bundhs

Improving dry bundhs can be accomplished by adhering to several key considerations:

- Carefully choosing shallow sloping depressions and undulating terrain composed of sandy soils with ample catchment areas.
- Building a small earthen bundh at the opposite end of the depression, away from the catchment area, to retain water for a specific duration. Maintaining a maximum water depth of 2 meters in the bundhs is crucial, with overflow water safeguarded using finely meshed wire netting.
- Creating spawning grounds at various levels in shallow bundhs to facilitate the breeding of major carp. Ensuring flooding at different water levels within the bundh is essential, positioning the spawning ground away from the current direction.
- Providing adjacent storage facilities such as storage tanks, cement cisterns, or earthen ponds to house breeders temporarily before their introduction into the bundh.
- Establishing a series of 10-20 rectangular cement hatcheries, each measuring 2.4 x 1.2 x 0.3m, to support the breeding process.
- A small double-storied building will be erected as an observation tower, storage facility, and shelter for workers involved in bundh management (Sakhare, 2007).

The problem in bundh Breeding

- Coordinating the simultaneous collection and incubation of large quantities of eggs, particularly in wet bundh breeding, presents considerable challenges.
- During the egg collection process in wet bundhs, it is expected to inadvertently gather unwanted fish spawns, predatory insect larvae, and other undesirable organisms.
- Despite achieving high fertilization rates, the hatching success and survival of hatchlings until the spawn stage are frequently low. The implementation of modern hatchery techniques can aid in improving these outcomes.
- Many fairy shrimps, such as *Streptocephalus* and *Branchinella* sp., may be present in dry bundhs, especially during late breeding with three weeks

of water accumulation. They can control their population by applying bleaching powder at a rate of 1 ppm on the first day of water accumulation.

- Dry bundhs are primarily owned by the government and serve as water sources and irrigation. Consequently, fish breeding in these bundhs is considered a secondary activity, which complicates the regulation of water inflow and outflow for fishery purposes.
- Brood fish for dry bundh breeding are predominantly captured from wild habitats using gillnets or cast nets, which can cause injuries to the brood fish. Additionally, brood fish may carry infections or injuries that could negatively impact Breeding.
- Heavy rains after spawning can result in a substantial influx of water, leading to mechanical damage and destruction of gonadal products.
- Not enough emphasis is placed on monitoring water quality, including dissolved gases, harmful substances, and predatory creatures, either before releasing brood fish or throughout spawning and spawn development.
- During the late monsoon period, when water accumulates more, some dry bundhs might overflow, increasing the risk of losing seeds from the bundh.
- In the months following the monsoon, young fish become susceptible to bird predation as water levels drop.
- In specific dry bundhs with uniformly flat basins, gathering seeds becomes difficult as water levels decrease. Mortality can result from rising temperatures and increased turbidity in shallow water due to repeated netting operations.
- Delayed harvesting of fish seeds with reduced water levels worsens the issue of illegal fishing.
- Early major carp have high feeding rates. Inadequate food availability can trigger cannibalism, mainly if there is a noticeable difference in size groups, especially when brood fish are released in batches.
- Frequently, when a dry bundh harbours many fish seeds, water is drained for agricultural purposes, resulting in substantial stock loss from the dry bundh.
- Typically, spawn is left unattended in the dry bundh under natural conditions. Excessive silt, predatory insects, and copepods can cause extensive damage to developing eggs and juvenile fishes.
- During spawning, water levels may drop to critical levels, exposing many eggs in the peripheral areas of the dry bundh, leading to significant spawn mortality.

Induced breeding Technology

Carps, typically reproduce in free-flowing bodies of water such as rivers. They do not naturally breed in confined or restricted waters. When gathering carp seed (eggs or larvae) from the wild, the resulting stock typically consists of various types, both desirable and undesirable. Sorting out the desired carp seed from this mixed stock poses considerable difficulty. Handling the seed during this process may lead to the demise of the desired varieties.

Moreover, the presence of predatory fish seeds can pose a threat to the desirable carp seed. Another obstacle in obtaining the desired carp seed is the inability to collect a sufficient number from natural sources. Pure and untainted seed availability is scarce, further complicating the situation. Nevertheless, in tackling these challenges, induced breeding emerges as a highly effective approach for acquiring high-quality carp offspring that meet specific standards. This technique presents numerous benefits compared to exclusively depending on natural harvesting methods.

Induced breeding is a method that allows for producing pure seeds of desirable species. For example, if rohu seed is needed, it can be obtained within a couple of days using this technique. This approach enables the production of the required number of seeds efficiently and quickly. Suppose a fish farm requires one crore fish seeds; induced Breeding can produce this quantity soon in a shorter timeframe. Additionally, this method eliminates the issues of seed identification and segregation since only the desired species' seeds are produced. The technique itself is straightforward to implement. It ensures the production of healthy seeds and enables fish to spawn multiple times within a year.

Moreover, induced breeding facilitates hybridization, allowing for the creation of hybrid fish species. In induced breeding methods, fish are injected with four primary substances: extracts from the pituitary gland, HCG (human chorionic gonadotropin), ovaprim, and ovatide. These substances play a crucial role in stimulating fish to undergo spawning and facilitate the collection of eggs and milt for breeding purposes (Felix *et al.*, 2020).

Induced breeding with pituitary gland extraction

Fish breeding using pituitary gland extraction is reliable for acquiring pure fish seed. It is widely practised in India and several other countries. This technique entails injecting mature female and male fish with pituitary gland extracts from other mature fish.

Historical background

To spawn by injecting them with pituitary extracts. This breakthrough marked a crucial step in understanding the pituitary gland's role in vertebrate reproduction.

The pituitary gland, the master gland, is pivotal in regulating various physiological processes, including growth, metabolism, and reproduction. Aschheim and Zondek (1927) demonstrated that introducing pituitary implants into female mice could accelerate their sexual maturation. This discovery sparked additional inquiries into the reproductive impacts of the pituitary gland. Subsequently, in 1930, Houssay from Argentina pioneered similar experimentation on fish, marking the first instance of such research in aquatic organisms. Houssay triggered the premature birth of developing young by injecting extracts from the pituitary gland of one fish species into a viviparous catfish. This demonstrated the influence of pituitary gland extracts on fish reproduction, expanding the scope of research beyond mammals. Building upon these earlier experiments, Von Ihring 1934 devised a successful technique for inducing spawning in certain Brazilian pond fishes. He accomplished this by administering a solution containing freshly harvested pituitary glands from less significant fish species. This innovation enabled Brazilians to utilize hypophysation, artificially stimulating fish breeding by targeting the pituitary gland. Subsequently, in 1937, Russian scientist Gerbiskii achieved a significant milestone by successfully inducing sturgeons, specifically *Acipenser stellatus*, to spawn through pituitary extract injections. This accomplishment marked a notable advancement in understanding the reproductive mechanisms of higher vertebrates (Grandi, 2014).

India is the third country to incorporate the technique into its fish farming program. In 1937, an initial endeavour in India was made by Hamid Khan, who attempted to stimulate spawning in *Cirrhinus mrigala* by injecting pituitary gland extracts from mammals. Following this, in 1945, Hussain experimented with hormones such as 80-120 RV Prolan and Antuitrin-S, aiming to induce spawning in female *Labeo rohita* and *Cirrhinus mrigala*. Subsequently, in 1955, Hiralal Choudhuri achieved a breakthrough by triggering spawning in *Esomus detritus through intraperitoneal* injection of *a* pituitary extract derived from *Catla catla*. Additionally, he successfully bred *Pseudotropius atherinoides*. Later, in 1955 and 1956, Ramaswamy and Sunderaraj managed to breed *Heteropneustes fossilis* and *Clarias batrachus*. The first successful induced breeding of Indian major carp using hypophysation occurred in 1957, and it was accomplished by Hiralal Chaudhuri and Alikunhi at CIFRI, Cuttack.

Pituitary gland

The fish pituitary gland is a small, soft body that appears creamish white. In carp, it is more or less round. Located on the brain's underside, it sits behind the optic chiasma within a depression in the base of the skull called the 'Sella turcica'. Surrounding the gland is a delicate membrane known as the dura mater. In some types of fish, a slender structure called the infundibular stalk is linked to the brain.

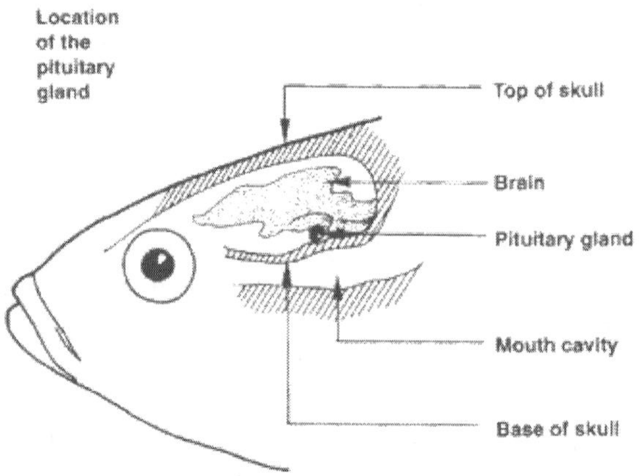

Picture 3: Location of pituitary gland (https://www.fao.org/fishery/docs)

The pituitary glands are categorized into two types: platybasic, which lacks a stalk, or leptobasic, which possesses a stalk, based on whether the infundibular stalk is present. The platybasic glands lack a stalk and possess an open infundibular recess, while the leptobasic glands have a stalk and an obliterated infundibular recess. Carps typically have the leptobasic type of pituitary glands, while Channidae and Nandidae fish have the platybasic type. The size and weight of the pituitary gland vary by the size and weight of the fish it belongs to. For instance, in *Labeo rohita*, the pituitary gland weighs approximately 6.6 mg in fish weighing 1-2 kg, 10.3 mg in fish weighing 2-3 kg, 15.2 mg in fish weighing 3-4 kg, and 18.6 mg in fish weighing 4-5 kg (Sharif *et al.*, 2022).

Responsible for secreting gonadotropic hormones—specifically FSH (Follicular stimulating hormone) and LH (Luteinizing hormone)—the pituitary gland ensures consistent secretion throughout the year, with levels directly correlated to gonadal maturity cycles. FSH prompts ovarian follicle growth and maturation in females and supports male spermatogenesis. Conversely, LH facilitates the transition of ovarian follicles into corpus luteum in females and stimulates testosterone production in males. Notably, these hormones transcend species boundaries, meaning a hormone from one species can stimulate gonads in another species, albeit with varying effectiveness. In experiments aimed at inducing fish breeding, fish pituitary extracts have proven more effective than pituitary hormones, sex hormones, and mammalian-derived steroids. Consequently, fish pituitary extracts are widely utilized in global fish breeding endeavours (Beshav & Carr, 2017).

Collection of the pituitary gland

The fish that donates its pituitary gland is the donor fish. The success of fish breeding hinges upon the careful selection of the appropriate donor fish. The gland must be harvested from fully mature gravid fishes before they spawn, as it exhibits its maximum potency during this period. Glands from immature or spent fish are usually ineffective. Glands from induced-bred fishes can breed other fishes immediately after spawning. In India, the best time to collect pituitary glands of major carp is from May to July, when they reach advanced maturity. As a perennial breeder, common carp can provide mature individuals throughout the year for gland collection. "Glands are usually collected from freshly killed fishes, but those from ice-preserved specimens are also usable. Several techniques are used for pituitary gland collection worldwide.

In India, the standard method involves removing the scalp of the fish skull with a butcher's knife. After scalp removal, the brain's grey matter and fatty substances are gently cleaned using cotton. The brain is then detached from the nerves and carefully lifted out. In most cyprinids, the gland remains on the floor of the brain box after the brain is lifted. Carefully, the dura mater covering the gland is removed using precise tools like a fine needle and forceps. The gland is then delicately picked up to prevent harm, as damaged or broken glands tend to lose effectiveness (Shanthanagouda & Khairnar, 2018).

Another method of gland collection involves using the foramen magnum. Professionals in crowded fish markets commonly use this method for mass-scale collection. In this method, the fish must be beheaded. In the market, glands are gathered from the remnants of fish heads that have been pre-cut. The opening at the rear, which contains a mixture of grey matter and fatty substances, is readily identifiable. Positioned beneath this opening is the brain. To access the gland, the grey matter and fatty substances are initially removed by inserting the blunt end of forceps into the opening and extracting the entire material while ensuring the brain remains undisturbed. With caution, the brain is lifted, pushed forward, or pulled out from the opening. Subsequently, the gland on the brain cavity's floor is delicately retrieved using fine tweezers. Employing this collection method, an experimenter can conveniently amass approximately 50-60 glands within an hour (Woynarovich & Horyath, 1980).

Preservation of pituitary gland

If the glands gathered are not promptly utilized, they must be preserved because of their glyco- or muco-protein composition, rendering them vulnerable to enzymatic processes. There are three methods for preserving pituitary glands: absolute alcohol, acetone, and freezing. In India, the preferred method for preserving fish pituitary glands is absolute alcohol. Thus far, experiments with alcohol-preserved glands of Indian major carp have shown more favourable outcomes than those preserved with acetone. Following collection, the glands undergo defatting and dehydration in absolute alcohol. Each gland is then individually placed in a labelled vial for easy identification. After 24 hours, they are rinsed with absolute alcohol and transferred to fresh, dark-coloured bottles of absolute alcohol. These preserved glands can be stored at room temperature or in a refrigerator. Regularly replacing the alcohol solution helps maintain the quality of the glands over time. The vials can be placed within a desiccator containing anhydrous calcium chloride to prevent moisture ingress. Ideally, the glands should be stored in a refrigerator. They can be kept in the refrigerator for up to 2-3 years and at room temperature for up to one year (Linsler *et al.*, 2017).

Acetone is commonly used as a preservative for glands in the seed production process. After being collected, the glands are submerged in fresh acetone or dry ice-cooled acetone in a refrigerator set at -10°C for 36-48 hours. To ensure thorough removal of fat and moisture, the acetone is refreshed 2-3 times at intervals of around 8-12 hours. Once defatting and dehydration are completed, the glands are removed from the acetone and positioned on filter paper, allowing them to air dry at ambient temperature for approximately one hour. Afterwards, the dried glands are stored in a refrigerator at -10°C, preferably within a desiccator containing calcium chloride or other appropriate desiccants to uphold optimal preservation conditions. Preserving glands in acetone is predominantly followed in the USSR and the USA (Chacko *et al.*, 2003).

Preparation of pituitary gland extract

First, the glands that have been preserved undergo weighing, a crucial step in accurately determining the dosage according to the breeders' weight. This weighing can be performed either individually or collectively. For a more precise measurement, it is advisable to weigh the gland precisely two minutes after removing it from alcohol.

Preparing the pituitary extract is essential just before the injection. Initially, the amount of land needed for the injection is calculated based on the breeder's weight. Then, the suitable glands are selected, and the required quantity is extracted from the vials. If the glands have been preserved in alcohol, the alcohol evaporates. However, if the glands have been dried using acetone, they can be directly taken from the vials for maceration.

The glands are initially disintegrated into a pulp using a tissue homogenizer. This process involves adding a precise quantity of either distilled water, a saline solution, or a physiological solution with the same osmotic pressure as the recipient fish's blood. For Indian major carp, the most effective outcomes in induced Breeding have been observed with either distilled water or a 0.3% saline solution. Typically, the concentration of the resulting extract is maintained within the range of 1-4 milligrams of gland per 0.1 millilitres of the solution, roughly equivalent to 20-30 grams of gland per 1.0 millilitre of the solution. Once the homogenization is complete, the resulting suspension is transferred into a centrifuge tube. During the transfer, it is important to thoroughly mix the homogenate to ensure that any settled gland particles become suspended in the solution and make their way into the centrifuge tube.

The material inside the centrifuge tube and the extract undergoes centrifugation to separate heavier particles, which settle at the bottom. The clear liquid above these particles, known as the supernatant fluid, is then delicately collected using a hypodermic syringe for injection. Another approach involves preparing the pituitary extract in more significant amounts and preserving it in a glycerine solution, with a ratio of one part extract to two parts glycerine. This allows avoiding the inconvenience of preparing the extract each time before injection, particularly during the fish breeding season. It is crucial to store the stock extract in a refrigerator or on ice to maintain its viability (Shanthanagouda & Khairnar, 2018).

The technique of breeding

Induced Breeding for major carp usually occurs at the beginning of the usual monsoon period, coinciding with the fish's full maturity and the decrease in water temperature. This period is characterized by selecting female carp exhibiting a round, soft, and bulging abdomen, accompanied by a swollen reddish vent, as breeding candidates. Likewise, male carps showing freely flowing milt are chosen as breeders. Male breeders are distinguishable by the roughness on the dorsal surface of their pectoral fins.

Dosage of pituitary extract

In the process of inducing fish breeding, a critical consideration involves determining the appropriate quantities of pituitary extract. The efficacy of this extract hinges on several factors, including the size and developmental stage of the donor fish's sexual organs, the donor species, the timing of gland extraction, and proper preservation techniques. Dosage calculation typically involves measuring the extract relative to the recipient fish's weight. However, it is worth noting that even with identical doses administered to fish of similar weights, outcomes can vary due to differences in gonadal maturity. In instances where gonads are undergoing resorption, even high hormone doses may not produce the desired results. Successful Breeding entails meticulous selection of breeders and administering a specific dosage of pituitary gland extract per kilogram of their body weight.

Studies focusing on establishing consistent dosages have shown that giving female breeders a small initial dose, followed by a higher effective dose six hours later, yields better results than a single large dose. However, a single large dose has proven effective under optimal conditions and favourable weather. Rohu Seed Production technologies have displayed favorable outcomes with two injections, while catla and mrigal have responded well to single and double injections.

The breeding process commences by administering an initial dosage of 2-3 mg of pituitary gland per kilogram of body weight in female breeders. This initial dosage is not necessary for male breeders unless they release milk under slight pressure on their abdomen. Each breeding set comprises two males paired with a female, ensuring their combined weight equals or surpasses the females for a good match. If one male is not in the freely oozing stage, an initial injection of 2-3 mg per kilogram of body weight may be administered. Following a 6-hour interval, the female receives a second dosage of 5-8 mg per kilogram of body weight. Concurrently, both males receive either the first or second dose at a rate of 2-3 mg per kilogram of body weight. Notably, minor dosage adjustments may be warranted based on the breeders' maturity and environmental conditions. In cases of chemical imbalance, effectively stimulating a pair of fish entails using 1-3 pituitary glands (Tiwana & Raman, 2012).

Methods of injection

In the USSR, intracranial injections were favoured, whereas in the USA and Japan, intraperitoneal injections were preferred. In India, intramuscular injections are most commonly administered. This method is perceived as less hazardous compared to others.

Intraperitoneal injections are usually given in the soft areas of the body, often at the base of the pelvic fin or occasionally at the base of the pectoral fin. However, there is a potential risk of harming internal organs, particularly the distended gonads, when administering an intra-peritoneal injection to mature fish fully.

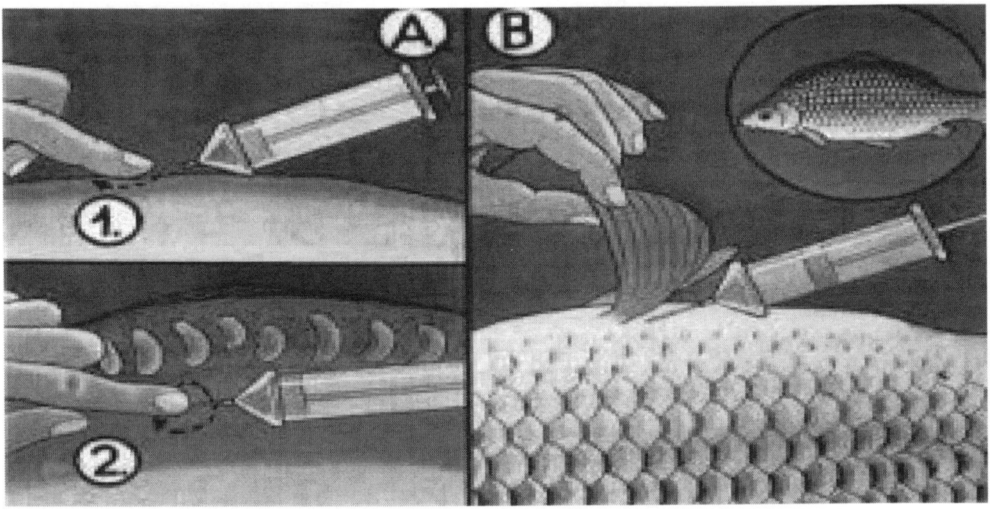

Picture 4: (Intraperitoneal injection (https://www.fao.org)

Injections are commonly administered at the caudal peduncle or the shoulder areas close to the base of the dorsal fin in carps. When administering injections to carps, the needle is placed beneath a scale, parallel to the fish's body, and inserted into the muscle at an angle. There are no strict guidelines regarding the timing of injections; they can be administered at any time of the day or night. However, because lower temperatures are beneficial and the nighttime tends to be quieter, injections are typically given during the late afternoon or evening hours. This timing allows the fish to capitalize on the tranquillity of the night for uninterrupted spawning.

The most suitable hypodermic syringe for this purpose is a two cc syringe with markings at 0.1 cc intervals. Needle selection depends on the size of the carp being injected: for carps weighing 1-3 kg, a No. 22 needle is typically employed, whereas more enormous carps necessitate a No. 19 needle. More minor carps can be injected using a No. 24 needle.

Anaesthetics are advised to enhance the survival of brood fish during injection. Two commonly utilized anaesthetics are MS 222 and Quinaldine. MS 222 can be diluted in water at 50-100 mg per litre. Alternatively, a cotton roll soaked in a 0.04 M solution of MS 222 can be placed inside the fish's mouth. Quinaldine is administered at 50-100 mg per litre (Surnar *et al.*, 2015).

Breeding hapa and spawning

After administering the injection, the breeders are placed inside the designated breeding enclosure, a breeding hapa. These hapas are typically crafted from delicate cloth and come in varying dimensions, with more prominent breeders housed in structures measuring 3.5 x 1.5 x 1.0 m, while more miniature breeders weighing less than 3 kg occupy units sized at 2.5 x 1.2 x 1.0 m. Each breeding hapa is fully enclosed, with all sides securely stitched shut, except for a top opening for introducing the breeders. Generally, one pair of breeders is introduced into each hapa. However, in certain instances, communal breeding practices are adopted to minimize the number of male breeders required and to economize on pituitary material. Once the breeders are released into the hapa, the opening is tightly sealed to prevent any escape attempts. Alternatively, cement cisterns or plastic pools of similar dimensions to the hapas may be utilized for breeding purposes.

Spawning typically takes place within 3-6 hours after the second injection. Following fertilization, the eggs undergo significant swelling due to water absorption. Fertilized eggs of major carps exhibit a shining, glass-bead-like appearance with crystal-clear transparency, while unfertilized eggs appear opaque and whitish. The size of eggs can vary considerably among breeders of the same species. Fully swollen eggs of Indian major carp have a diameter of 2.5 mm, with catla having the most giant eggs and rohu having the smallest. The great Indian carp, masters of reproduction, boast prolificacy: rohu averaging 3.1 lahks, catla varying from 1 to 3 lakh, and mrigal reaching 1.5 lahks in fecundity.

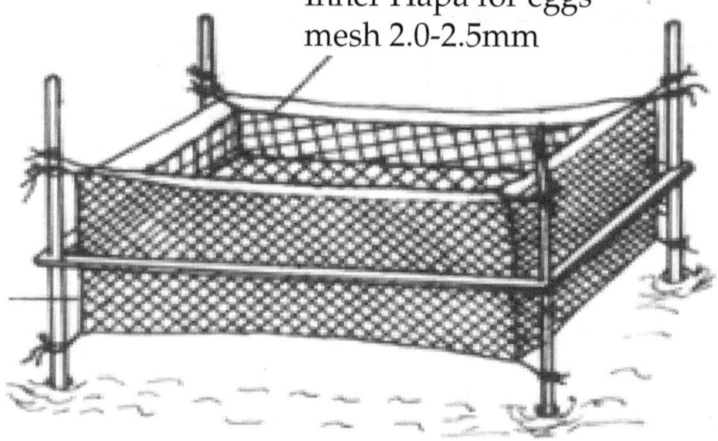

Picture: 5 Hatching Hapa (Surnar *et al.*, 2015)
(https://images.app.goo.gl/eDENYKKbw5q4F4sD6)

During the breeding process, the developing eggs go through a critical phase inside the breeding enclosure, where they remain untouched for at least 4-5 hours post-spawning. This period allows the eggs to undergo proper conditioning in the water. Afterwards, the eggs are delicately collected from the enclosure using a container and then transferred into a bucket with a small amount of water. The breeders are initially removed from the enclosure and weighed to estimate the egg quantity. The variance in weight before and after spawning indicates the approximate number of eggs laid.

Furthermore, the eggs total volume and count can be easily determined based on the known volume and count within the sample container. Random sampling is conducted pre- and post-spawning to ascertain the percentage of fertilized eggs. This evaluation allows for an estimation of the proportion of eggs that have been successfully fertilized (Surnar *et al.*, 2015).

Stripping

On the other hand, Chinese carp do not reproduce naturally, and even when they do, the fertilization rate tends to be relatively minimal. Consequently, the process of stripping or artificial insemination is employed. The female fish is carefully positioned with its head inclined upward, tail down, and belly facing the receptacle. The eggs are gathered into an enamel or plastic trough by gently applying pressure to the female's body. Similarly, the male fish is held, and the milt is extracted and added to the same trough. To facilitate fertilization, the gametes are promptly mixed using a quill feather. Afterwards, the fertilized eggs are rinsed with clean water to eliminate excess milt. After being left alone in freshwater for about half an hour, the eggs are readied for placement into the hatching tanks (Szabo *et al.*, 2019).

Picture: 6 Stripping
(https://rb.gy/1ohx4i)

The technique of hatching the eggs:

The eggs obtained from breeding hapas are transferred to the hatching hapas. The hatching hapa consists of two distinct sections: the outer and inner hapa (refer to Figure 3.3). The inner hapa is smaller in size and fits inside the outer hapa. The outer hapa is made of thin cloth with 2 x 1 x 1 m dimensions, while the inner hapa is made of round meshed mosquito net cloth measuring 1.75 x 0.75 x 0.5 m. Loops and ropes are provided at all corners of the outer and inner hapas to facilitate installation.

In hapa-based aquaculture, each inner hapa harbours a considerable number of eggs, ranging from 75,000 to 100,000, which are uniformly dispersed. These eggs undergo hatching within 14 to 20 hours, contingent upon a temperature spectrum of 24° to 31°C. Notably, the incubation duration exhibits an inverse correlation with the prevailing temperature conditions. Upon hatching, the nascent hatchlings navigate through the apertures of the inner hapa, transitioning into the outer enclosure. The inner hapa, having served its purpose, retains remnants such as empty egg casings and non-viable eggs, necessitating a cleaning regimen upon the culmination of the hatching phase.

After emergence into the outer hapa, the hatchlings experience an uninterrupted phase until the third day post-hatching. Throughout this interval, their sustenance derives from the reserves stored within the yolk sacs. On the third day, a pivotal developmental milestone is reached as the hatchlings manifest a discernible mouth and initiate locomotion, concurrently engaging in feeding activities. At this juncture, meticulous retrieval of the hatchlings from the outer hatching hapa is executed, segueing into their relocation to meticulously prepared nurseries. These nurseries serve as pivotal environments for the continuation of their growth and development.

Recent research has revealed that Indian major carp have the potential to undergo two spawning cycles within a single season, separated by a two-month interval. After the first spawning event, the breeders are given a diet comprising groundnut oilcake and rice bran in equal proportions, making up 2.5% of their body weight. As conducive climatic conditions emerge, these breeders mature and are primed for the following spawning phase (Mukherjee & Praharaj, 2002).

Induced breeding with ovaprim

Due to inconsistent pituitary potency, alternatives were explored. LH-RH analogues were used in fish breeding attempts worldwide with varying success. However, LH-RH had inconsistent results and required higher doses for spawning induction. A joint effort between Dr. Lin from China and Dr. Peter from Canada, supported by the International Development Research Centre, led to the creation of a dependable breeding method known as the 'LNPE' approach. This method integrates an LH-RH analogue with a dopamine antagonist. Following this concept, Syndel Laboratories Limited in Canada produced a novel medication named 'ovaprim'.

Ovaprim remains stable at room temperature and is prepared for immediate utilization. It comprises an LH-RH analogue, equivalent to 20 μg of Salmon gonadotropin-releasing hormone (GnRH), and a dopamine antagonist called domperidone at a 10 mg/ml concentration. Ovaprim consistently demonstrates potency, with sGnRHa being recognized as 17 times more potent than LH-RH (Peter, 1987). The inclusion of the dopamine antagonist domperidone in ovaprim is considered superior to the commonly used antagonist, pimozide. Ovaprim is a convenient and efficient ovulation inducer, as it is readily available for use and does not require refrigeration for storage.

This medication is administered to male and female brood fish simultaneously in a single dosage, unlike pituitary extract, which requires two separate doses. This approach minimizes the handling of brood fish, saving considerable time and labour and contributing to the overall cost efficiency of seed production. The spawning response observed in treated species surpasses those injected with pituitary extract. Ovaprim has exhibited highly encouraging outcomes in stimulating the Breeding of various carp species, including catla, rohu, mrigal, silver carp, grass carp, and oversized heads. The necessary dosage varies across different carp species. Generally, the recommended dosage for all carp species is 0.10-0.20 ml ovaprim per kilogram of body weight for males and 0.25-0.80 ml ovaprim per kilogram of body weight for females. Female catla displays a positive response within the dosage range of 0.4-0.5 ml/kg, while rohu and mrigal require lower dosages of 0.35 ml/kg and 0.25 ml/kg, respectively. Among exotic carp, silver carp and grass carp are bred with doses ranging from 0.40-0.60 ml/kg. Big-head carp have shown successful breeding outcomes at 0.50 ml/kg dosage. Optimum dosages for male Indian carps range from 0.10-0.15 ml/kg, while for exotic male carps, it is 0.15-0.20 ml/kg. The injection method is consistent with that used for pituitary extract. Ovaprim is widely used for induced fish breeding in many countries. Initial trials were conducted in Karnataka, Andhra Pradesh, and Tamil Nadu in India 1988.

Ovaprim presents distinct benefits over pituitary hormones. It is conveniently available in a pre-mixed liquid state in a 10 ml vial, ensuring consistent strength and dependable outcomes. Its extended shelf life and capacity for room temperature storage make it particularly practical. The formulation guards against excessive dosing, enabling simultaneous injection for male and female fish with a single dose. This streamlined process minimizes handling and post-breeding loss. Moreover, it facilitates multiple spawning events later in the season and yields a notable rate of egg production, fertilization, and successful hatching, as indicated by (Ali *et al.*, 2015).

Induced breeding with Ovatide

Ovatide is a novel hormonal formulation explicitly designed to induce Breeding in fish. It offers several advantages, such as being indigenous, cost-effective, and highly effective in stimulating spawning. This new formulation is based on a synthetic peptide that shares a structural similarity with the naturally occurring hormone known as gonadotropin-releasing hormone (GnRH). Unlike steroidal hormones, GnRH is classified as a peptide and under organic substances.

Ovatide is presented as an injectable solution with low viscosity, making it easy to administer. Compared to other commercially available spawning agents, Ovatide not only demonstrates high activity but also proves to be more cost-effective. It has shown efficacy in breeding major carp and catfish, making it suitable for diverse fish species.

The recommended dosage for females varies depending on the fish species. For rohu and mrigal, the dose is 0.20-0.40 ml/kg, while for catla, silver carp, and grass carp, it is 0.40-0.50 ml/kg. Calbasu needs 0.20-0.30 ml/kg. Males need less: rohu, mrigal, and calbasu need 0.10-0.20 ml/kg, catla needs 0.20-0.30 ml/kg, and silver carp and grass carp need 0.20-0.25 ml/kg.

Ovatide has many benefits. It is cost-effective and results in high fertilization (85%-95%). It encourages complete spawning, increases egg production, and ensures healthy seed production. Its low viscosity makes it easy to inject without harming the brood fish. Ovatide simplifies Breeding by allowing a single dose for brooders. It can be stored at room temperature, offering convenience in storage. Even under unfavourable climatic conditions, ovatide remains highly influential. "It is sold in a 10 ml vial, costing Rs. 300, offering a more economical option than ovaprim. The process for choosing brooders and administering injections for ovatide resembles the methods utilized for pituitary extract, as outlined in Tiwana and Raman's study from 2012.

Induced Breeding with Ovopel:

Ovopel, a product developed by the University of Godollo in Hungary, is a formulation comprising mammalian GnRH (gonadotropin-releasing hormone) and metoclopramide, a water-soluble dopamine receptor antagonist. Each pellet of Ovopel contains approximately 18 to 20 micrograms of D-Ala6, Pro9NEt-mGnRH and 8 to 10 milligrams of metoclopramide. Administered in pellet form, Ovopel contains a potent analogue of the gonadotropin-releasing hypothalamic hormone, equivalent to the effect of a 3 mg normal acetone-dried dehydrated carp hypophysis gland. Research suggests that the effectiveness of inducing fish propagation is heightened when the hormone is administered in two stages: a prime dose followed by a resolving dose, as outlined by Szabo in 1996. Successful cyprinid outcomes were observed when female brood fish were given 2 to 2.5 pellets per kilogram. Additionally, preliminary trials involving a single injection of Ovopel have shown promising results with certain species of Indian major carps and *Clarias batrachus.*

The necessary quantity of ovopel was determined according to the weight and condition of the brood fish. After crushing, the pellets were dissolved in water. Trials conducted during July-August 1999 revealed that ovopel, as per Szabo it is convenient to store, simple to administer, and cost-effective. However, extensive research is essential in India to ascertain its efficacy and economic feasibility. In Europe, ovopel has been successfully trialled for inducing ovulation in various cyprinid species, including Common carp, Silver carp, and tench (Horvath *et al.*, 1997). Ovulation induction has also been observed in African Catfish (Brzuska, 1998). Successful trials employing ovopel for induced Breeding of major carp have been conducted in Uttar Pradesh, Haryana, and Punjab. Recent trials in Assam involving species like *Labeo rohita* (Rohu), *Cirrihinus mrigala* (Mrigal), *Labeo gonius* (Gonius), and *Clarias batrachus* (Magur) have demonstrated promising outcomes. This indicates the potential utility of this novel hormone preparation for commercial fish seed production, contingent upon its availability to farmers at a competitive price (Souza *et al.*, 2018).

Other compounds, such for induced Breeding

Other compounds, such as LH-RH analogues, steroids, and clomiphene, stimulate fish breeding artificially.

LH- RH analogues

Various substitutes for Luteinizing hormone-releasing hormone (LH-RH) have been utilized in fish breeding through artificial means. Extensive investigations have demonstrated that the simultaneous utilization of dopamine antagonists alongside these analogues yields promising results in China, with successful dosages ranging from 10 to 100 µg/kg. Interestingly, an analogue of teleost GNRH (Gonadotropin-releasing hormone) has exhibited greater potency than LH-RH. In teleosts, GnRH stimulates the production of Gonadotropin hormone (GTH), with recommended dosages ranging from 25 to 100 µg/kg (Hawarry *et al.*, 2016).

Steroids

Certain steroid hormones are utilized to stimulate fish reproduction. The main impact of these hormones on ovulation is seen in the breakdown of the germinal vesicle (GVBD). In mullets, ovulated eggs take at least 4 hours to become capable of fertilization, whereas in most other fish species, eggs are immediately fertilizable. It is known that pituitary gonadotropins act on oocyte maturation through mediation by steroid hormones. In Heteropneustes fossilis, Deoxycorticosterone acetate (DOCA) and cortisone have been shown to effectively stimulate ovulation (50 mg/kg of fish) (Goswamy & Sunderraj, 1971).

Additionally, the compound known as 17á-hydroxy-20B dihydroprogesterone (17á-20BDP) has proven effective in stimulating ovulation in goldfish, trout, and pikes, as noted by Jalabert in 1973. Other commonly employed steroid hormones for inducing spawning in teleosts encompass cortisone acetate, deoxycortisol, deoxycorticosterone, hydroxycortisone, progesterone, 11-deoxycorticosterone, and 20B progesterone. The benefits associated with the use of steroids are manifold: most of these compounds are readily available in synthetic forms as highly pure preparations, ensuring consistent quality, and they are generally more cost-effective compared to gonadotropin preparations, as highlighted (Hoga *et al.*, 2018).

Clomiphene

The synthetic non-steroidal estrogen chlorotrianisene has an analogue called clomiphene, which has antiestrogenic effects in fish. Clomiphene stimulates the release of gonadotropins and has been found to induce ovulation in goldfish within four days when injected at a dose of 10 µg/g. In common carp, successful spawning occurred after 40-64 hours when the exact dosage of clomiphene was administered (Kapur, 1981).

Estimation of eggs

Eggs gathered from a " hapa " breeding device using cups, trays, or beakers are then moved into buckets. During this process, breeders are also separated, and their weights are recorded. The variance in weight serves as an estimate for the number of eggs that have been laid. Eggs are kept in a meshed net, and water is drained. They are measured in known volume containers (beaker/mug/cup) and transferred to hatcheries. The total quantity is estimated from the measured volume. The fertilization percentage was calculated by counting fertilized eggs from a 1 ml egg sample.

Common carp breeding

The common carp (*Cyprinus carpio*) usually breeds in restricted aquatic habitats. Reproduction occurs in shallow regions with marginal plant life, frequently overrun by aquatic vegetation. This process occurs predominantly during two primary intervals: spanning from January to March and July to August. However, it has been observed that common carp can breed throughout the year. Controlled breeding practices are employed to enhance the success of spawning and hatching. To commence controlled Breeding, a designated set of parent fish comprising a single female and two males are introduced into a breeding hapa, a net enclosure. Occasionally, to promote successful spawning, the female fish might receive a small injection of pituitary gland extract, usually in the range of 2 to 3 mg per kg of body weight.

Freshly washed aquatic weeds such as *Hydrilla, Najas, Eichhornia,* and others are evenly distributed inside the hapa. These aquatic weeds serve as collections for the fish eggs. The weed used is approximately twice the weight of the female fish introduced. Each weed carries around 40,000 to 100,000 eggs, distributed within a single hatching hapa. After 4 or 5 days, the weeds are carefully removed from the hapa, taking care not to disturb the developing eggs (Chen *et al.*, 2022).

References

Agarwal, R. L. (2018). *Seed technology. Oxford and IBH Publishing.*

Ali, M. A., Rasheed, S. B., Hassan, Z., Ibrar, M., Majeed, A., Ulhaq, Z. & Khan, H. (2015). Efficacy of synthetic hormones Ovatide and Ovaprim in induced breeding of prominent Indian and Chinese carps. *Journal of Agricultural Technology, 11*(7), 1449–1456.

Bagchi, A. & Jha, P. (2011). Fish and fisheries are part of Indian heritage and pisciculture development. *Reviews in Fisheries Science, 19*(2), 85–118.

Beshay, V. E. & Carr, B. R. (2017). Hypothalamic–pituitary–ovarian axis and control of the menstrual cycle. *Clinical reproductive medicine and surgery: A practical guide*, 1–17.

Chacko, A. G., Chacko, G., Seshadri, M. S. & Chandy, M. J. (2003). The 'capsule ' of pituitary macroadenomas represents normal pituitary gland: a histopathological study. *British Journal of Neurosurgery, 17*(3), 213-218.

Chattopadhyay, N. R., Ghorai, P. P. & SK, D. (2013). Bundh breeding-rejuvenation of a novel technology for quality seed production by the fish seed producers of Bankura District in West Bengal, India. *International Journal of Current Science, 9*, 123–132.

Chen, L., Xu, J., Sun, X. & Xu, P. (2022). Research advances and future perspectives of genomics and genetic improvement in allotetraploid common carp. *Reviews in Aquaculture, 14*(2), 957-978.

De grandi, F. (2014). Determination of luteinizing hormone (LH) and follicle-stimulating hormone (FSH) in bovine plasma: development and validation of species-specific monoclonal antibody-based enzyme-linked immunosorbent assays (ELISA). https://hdl.handle.net/2434/246436

El-Hawarry, W. N., Abd El-Rahman, S. H. & Shourbela, R. M. (2016). Breeding response and larval quality of African catfish (*Clarias*) using different hormones/ hormonal analogues with dopamine antagonists. *The Egyptian Journal of Aquatic Research, 42*(2), 231–239.

Felix, S., Ahilan, B., Rajagopalasamy, C. B. T. & Moses, T. L. S. (2020). Induced Breeding and developmental biology of endemic Western Ghats Fish *Dawkinsia* filamentosa (Valenciennes, 1844) under captive conditions. *Indian Journal of Animal Research, 54*(9), 1069-1077.

Gangadhar, B., Sridhar, N., Hemaprasanth, K., Raghunath, M.& Jayasankar, P. (2016). Indigenous technical knowledge in aquaculture sector: A literature review. *International Journal of Fish Aquaculture Study*, 4(1), 373-378.

Hoga, C. A., Almeida, F. L.& Reyes, F. G. (2018). A review on the use of hormones in fish farming: Analytical methods to determine their residues. *CyTA-Journal of Food, 16*(**1**), 679-691.

Kapur, K. (1981). Spawning of major Indian carp—a review. *Acta hydrochimica et hydrobiologica*, 9(2), 163–182.

Linsler, S., Hero-Gross, R., Friesenhahn-Ochs, B., Sharif, S., Lammert, F. & Oertel, J. (2017). Preservation of hormonal function by identifying pituitary gland at endoscopic surgery. *Journal of Clinical Neuroscience*, 43, 240-246.

Mijkherjee, M., Praharaj, A. & Das, S. (2002). Conservation of endangered fish stocks through artificial propagation and larval rearing technique in West Bengal, India. *Aquaculture Asia*, 7(2), 8-11.

Routray, P., Verma, D. K., Sarkar, S. K. & Sarangi, N. (2007). Recent advances in carp seed production and milt cryopreservation. *Fish Physiology and Biochemistry, 33*, 413-427.

Routray, P., Verma, D. K., Sarkar, S. K. & Sarangi, N. (2007). Recent advances in carp seed production and milt cryopreservation. *Fish Physiology and Biochemistry, 33*, 413-427.

Sakhare, V. B. (2007). *Applied Fisheries*. Daya Publishing House. https://books.google.co.in/books/about/Applied_Fisheries.html?id=yJQJeKh7eM0C&redir_esc=y

Shanthanagouda, A. H. & Khairnar, S. O. (2018). Breeding and spawning of fishes:

Role of the endocrine gland. *International Journal of Fisheries and Aquatic Studies, 6*(4), 472–478.

Sharif, B. N., Ghosh, A., Mondal, S., Alam, M. H., Islam, M. A., Kabir, M. S. & Karim, M. M. (2022). Efficacy of dry and wet carp pituitary gland (CPG) in the induced Breeding of Cyprinus carpio var. specularis (Lacepède, 1803). *Asian Journal of Medical and Biological Research, 8*(2), 115-122.

Soliman, N. F. (2017). Aquaculture in Egypt under changing climate. *Alexandria Research Center for Adaptation to Climate Change (ARCA): Alexandria, Egypt.*

Souza, F. N., Martins, E. D. F. F., Corrêa Filho, R. A. C., de Abreu, J. S., Pires, L. B., Streit Jr, D. P. & Povh, J. A. (2018). Ovopel® and carp pituitary extract for induction of reproduction in *Colossoma macropomum* females. *Animal reproduction science, 195*, 53-57.

Surnar, S. R., Kamble, A. D., Walse, N. S., Sharma, O. P. & Saini, V. P. (2015). Hormone administration with induced spawning of Indian major carp. *International Journal of Fisheries and Aquatic Studies, 3*(1), 01–04.

Szabó, T., Urbányi, B., Müller, T., Szabó, R. and Horváth, L. (2019). Assessment of induced Breeding of major Chinese carp at a large-scale hatchery in Hungary. *Aquaculture Reports, 14*, 100193.

Tiwana, G. S. & Raman, S. (2012). An Economically Viable Approach for Induced Breeding of *Labeo Rohita* by Ovatide, Ovaprim And Carp Pituitary Extract. *IOSR Journal of Agriculture and Veterinary Science, 1*(1), 30-32.

Woynarovich, E. & Horváth, L. (1980). *The artificial propagation of warm-water finfishes: a manual for extension, 201.*

CHAPTER 5

CULTURABLE FISHES OF INDIA AND ABROAD

Introduction

Culturable fishes refer to fish species that can be raised and cultivated in controlled environments such as fish farms or aquaculture facilities. Choosing acceptable fish species for aquaculture, setting up proper rearing facilities, and maintaining their development and reproduction are generally involved in introducing culturable fishes. The "blue revolution," as described by (Neori *et al.*, 2007 Ahmed *et al.*, 2010; Simpson, 2011), refers to the rapid development of aquaculture. It is widely regarded as a crucial solution to the food production challenge. It is considered the most viable option to enhance global human nutrition and food security, as pointed out by (Allison, 2011) and (Béné *et al.*, 2016). With an average annual growth rate of 5.3% between 2001 and 2018, aquaculture is the fastest-growing food production industry in the world. Production surged 600% from 13 million tonnes in 1990 to 82 million in 2018 (FAO, 2020). Fish farming is a highly productive venture, with fish being a valuable source of easily digestible proteins rich in essential amino acids like lysine and methionine.

Additionally, they provide a wealth of essential minerals such as calcium, phosphorus, iron, sodium, potassium, magnesium, and sulfur, along with vitamins A and D and health-promoting fats. Polyunsaturated fatty acids in fish are especially beneficial for cholesterol regulation and promoting heart health. By recognizing the immense nutritional value and economic potential of fish farming, we can encourage its expansion and adoption, leading to a more sustainable and prosperous future.

Fish cultivation, also known as aquaculture, has a long and diverse historical background that dates back thousands of years. Raising fish for food and other purposes has evolved and spread across different cultures and regions. Here are some key milestones in the fish culture:

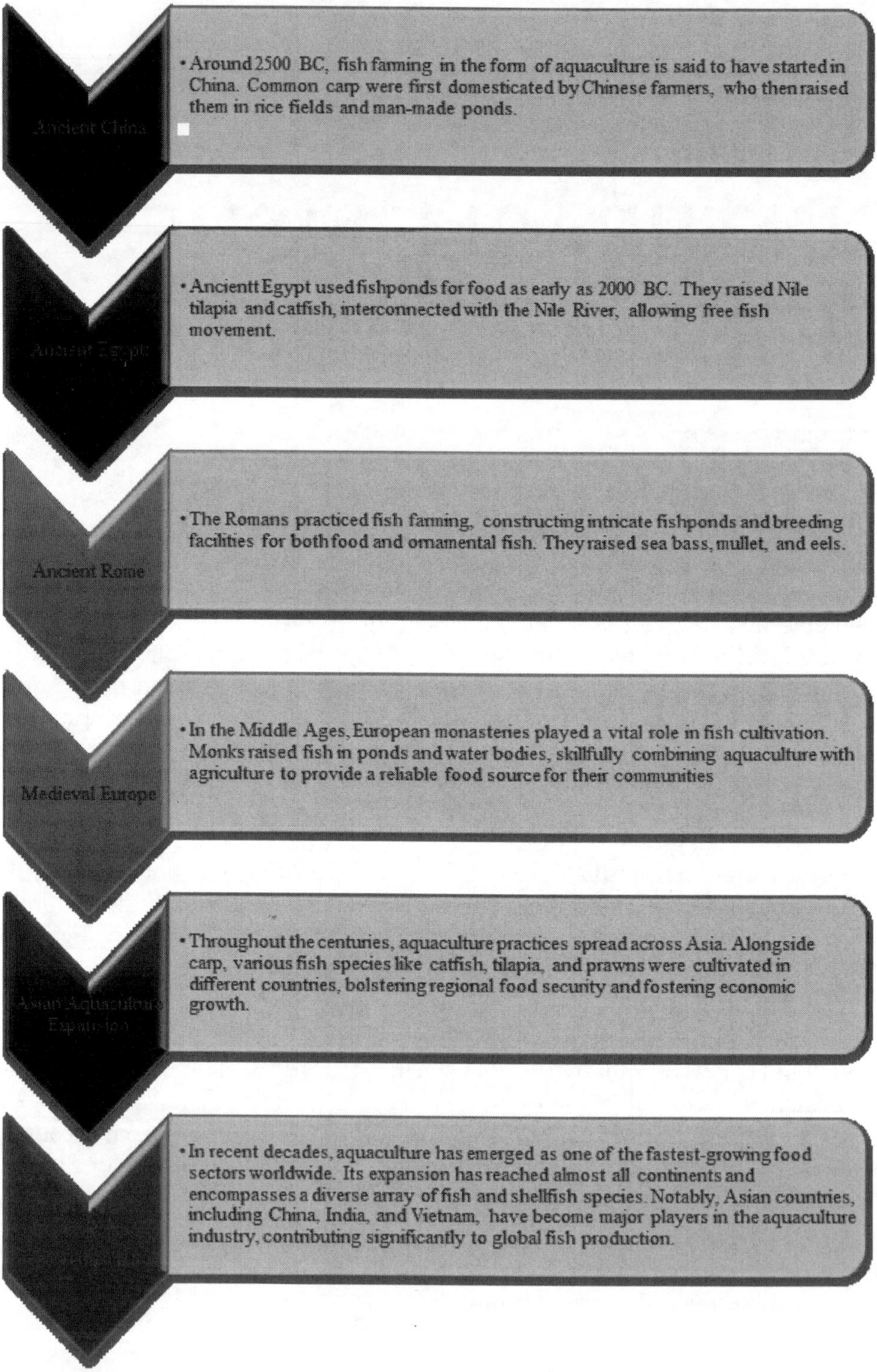
Ancient China
• Around 2500 BC, fish farming in the form of aquaculture is said to have started in China. Common carp were first domesticated by Chinese farmers, who then raised them in rice fields and man-made ponds.
Ancient Egypt
• Ancientt Egypt used fishponds for food as early as 2000 BC. They raised Nile tilapia and catfish, interconnected with the Nile River, allowing free fish movement.
Ancient Rome
• The Romans practiced fish farming, constructing intricate fishponds and breeding facilities for both food and ornamental fish. They raised sea bass, mullet, and eels.
Medieval Europe
• In the Middle Ages, European monasteries played a vital role in fish cultivation. Monks raised fish in ponds and water bodies, skillfully combining aquaculture with agriculture to provide a reliable food source for their communities
Asian Aquaculture Expansion
• Throughout the centuries, aquaculture practices spread across Asia. Alongside carp, various fish species like catfish, tilapia, and prawns were cultivated in different countries, bolstering regional food security and fostering economic growth.
• In recent decades, aquaculture has emerged as one of the fastest-growing food sectors worldwide. Its expansion has reached almost all continents and encompasses a diverse array of fish and shellfish species. Notably, Asian countries, including China, India, and Vietnam, have become major players in the aquaculture industry, contributing significantly to global fish production.

Culturable fishes play a vital role in aquaculture, substantially contributing to food security, economic progress, and environmental sustainability. Several important reasons underscore the significance of culturable fishes in this domain:

Food security: Culturable fish are crucial in ensuring food security as the world's population grows. With an increasing demand for nutritious and affordable protein sources, these fish provide a valuable option to supplement or replace traditional livestock production. Aquaculture helps meet the rising need for fish products while alleviating pressure on wild fish stocks, many already overexploited.

Economic benefits: The aquaculture of culturable fishes is a significant economic activity in numerous regions, supporting the livelihoods of millions of people. Fish farming generates employment opportunities in various sectors, including production, processing, distribution, and marketing, making valuable contributions to local and national economies.

Increased fish supply: Aquaculture enables controlled and efficient fish production year-round, ensuring a consistent and predictable fish supply. This aspect becomes particularly vital in regions with limited access to marine resources or where fish availability fluctuates seasonally.

Reduced pressure on wild fish stocks: There is substantial global concern over overfishing, which has led to the depletion of numerous wild fish populations. Through farming culturable fishes, aquaculture is critical in alleviating pressure on these wild fish stocks, allowing them to recover and preserve their ecological balance.

Environmental sustainability: Sustainable aquaculture practices strive to minimize adverse environmental impacts. By implementing well-managed fish farming systems, resources are efficiently utilized, pollution is reduced, and habitat destruction is minimized. Moreover, innovative techniques like integrated multi-trophic aquaculture (IMTA) use waste from one species as nutrients for another, fostering more eco-friendly and sustainable systems.

Diversification of diets: Culturable fishes provide diverse species with unique nutritional profiles. This diversity allows for dietary diversification, promoting improved nutrition and consumer health benefits.

Innovation and technological advancements: Aquaculture serves as a catalyst for research and technological progress in various domains, including fish nutrition, genetics, disease management, and environmental sustainability. These advancements not only enhance fish farming practices but also have far-reaching applications within aquaculture and beyond

Climate resilience: Certain culturable fish species display resilience to climate variations and can be farmed in regions vulnerable to extreme weather events or limited water resources. Their adaptability makes them a crucial element of climate-resilient agriculture and food production.

Biodiversity conservation: In certain instances, aquaculture efforts concentrate on culturing endangered or threatened fish species for conservation purposes. This approach safeguards the genetic diversity of these species and holds the potential for reintroducing them into their natural habitats.

Picture 1: Significance of fish farming for food security (Bovenkerk & Meijboom, 2012)

Culturable fish species in India

Picture 2: *Labeo rohita*
(https://www.researchgate.net)

Rohu is a freshwater fish from the Cyprinidae family. It is commonly found in rivers, lakes, and ponds across the Indian subcontinent and neighbouring regions. Rohu is one of South Asia's most essential and economically significant fish species, particularly in countries like India, Bangladesh, Nepal, and Pakistan (Ghosh *et al.*, 2020). The rohu is a freshwater species in South Asia extensively cultivated through aquaculture. It does not naturally breed in lake ecosystems, necessitating induced spawning techniques. Additionally, the rohu is highly valued as a sought-after game fish.

Catla (*Labeo catla*)

Catla, commonly known as the major South Asian carp, holds significant economic importance as a freshwater fish belonging to the carp family Cyprinidae. Its native habitat includes rivers and lakes in northern India, Bangladesh, Myanmar, Nepal, and Pakistan. However, due to its valuable traits, Catla has been introduced and extensively farmed in other South Asian regions. This species plays a pivotal role in aquaculture practices in the area, making it one of South Asia's most crucially cultivated freshwater fishes. It is often grown in polyculture ponds alongside other carp-like fish, most notably the Rohu labeo (*Labeo rohita*) and mrigal carp. Polyculture ponds promote a synergistic relationship among the different fish species, enhancing overall productivity and ensuring balanced growth.

Catla has a distinctive appearance is characterized by its large, deep-bodied, and elongated shape. It typically has a bright silver colour on its sides and belly, with a bluish-grey back. The mouth is terminal and upturned, which indicates its filter-feeding nature. These are plankton feeders; they feed on plankton and detritus material (Hora & Pillay, 1962). The breeding season for *Catla* usually occurs during the monsoon months, from June to September, when the water levels rise, and the fish have better access to flooded areas for spawning. They are oviparous, laying eggs that hatch outside the mother's body.

Picture: 3 Catla (*Labeo catla*) https://shorturl.at/LS9oc

Common carp (*Cyprinus carpio*)

Common carp is a freshwater fish native to Asia and Europe (Yaqoob, 2021). It was introduced to various regions worldwide for aquaculture and sport fishing purposes, becoming widespread and now found in many countries worldwide.

Common carp have a distinctive appearance: a large, robust body and thick scales. They can vary in colour, ranging from bronze to dark brown, and may have a golden or reddish tinge. The mouth of the common carp is characterized by two barbels, which are fleshy, whisker-like sensory organs.

These can thrive in aquatic habitats, including lakes, rivers, ponds, and slow-moving or still waters. They prefer areas with muddy or sandy bottoms and are often found near aquatic vegetation, as they feed on plants and detritus. (García-Berthou, 2001).These fish are omnivorous and opportunistic feeders. While they primarily consume plants, including aquatic vegetation, algae, and debris, they are also known to eat insects, small crustaceans, and molluscs.

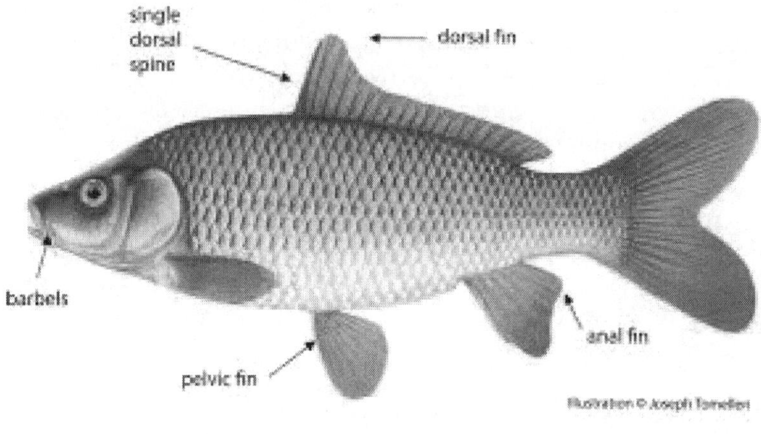

Picture: 4 Common carp
(https://www.asiancarp.carp)

Common carp are popular among anglers and have been introduced to many regions for recreational fishing. They are known for their strength and can offer a challenging fishing experience. Additionally, carp are widely raised in aquaculture facilities for their meat, especially in parts of Asia and Europe. Carp farming has a long history and plays a significant role in global aquaculture production. Despite their popularity for fishing and aquaculture, common carp can become invasive in non-native regions, where they can disrupt local ecosystems and cause environmental problems. Their feeding behaviour, especially uprooting vegetation, and their ability to reproduce rapidly can negatively impact native species and water quality. As a result, introducing and managing common carp in new areas requires careful consideration and control measures to prevent unintended consequences.

Mrigal (*Cirrhinus cirrhosus*)

Mrigal is the third major carp and is popular in fish farming due to its adaptability to various environmental conditions. It has a silver body with a slightly darker back, notable for its distinctive large scales. Mrigal is an omnivorous species that feeds on detritus, plankton, and aquatic plants.

Picture: 5 Mirgal (*Cirrhinus cirrhosus*)
https://rb.gy/xepne3

Silver carp is a freshwater fish species belonging to Cyprinidae family (Stepien *et al.*, 2019). It is native to eastern Asia, specifically found in rivers and large water bodies in China. Like other Asian carp species, silver carp has been introduced to various regions worldwide, including the United States, for aquaculture and to control algae and plankton in waterways. Silver carp have a distinctive, silvery appearance with a high-arched back and a laterally compressed body. They can grow quite large, reaching length of over three feet (1 meter) and weighing up to 60 pounds (27 kilograms). One of the most notable features of silver carp is its filtering apparatus, a specialized structure in its gills known as gill rakers, which allows it to filter large amounts of plankton and other small particles from the water. In their native range, silver carp inhabit rivers and large lakes, and they are known for their ability to thrive in both slow-moving and fast-flowing water. They are also highly adapted to environments with high turbidity, which allows them to feed on suspended plankton efficiently.

Filter-feeding Behavior: Silver carp are primarily filter feeders, and their feeding behaviour can have significant ecological impacts. They consume large quantities of plankton, including algae and phytoplankton, which are essential food sources for many native fish species. This voracious feeding can disrupt the food web and negatively affect the abundance and diversity of native fish and other aquatic organisms.

Tilapia is a common name for several species of freshwater fish belonging to the cichlid family (Cichlidae). These species are native to various regions worldwide, including Africa, the Middle East, and South America. Due to their rapid growth, hardiness, and mild taste, they have become one of the most widely farmed and consumed fish globally. Some popular tilapia species include:

Nile tilapia (*Oreochromis niloticus*): Nile tilapia is one of the most widely cultured and economically important tilapia species. It is native to Africa but has been introduced to many countries for aquaculture. Nile tilapia can tolerate various environmental conditions, making it suitable for farming in different regions (Eknath *et. al.*, 2009).

Picture 6 : Nile tilapia
(https://wiki.fishingplanet.com/Nile_Tilapia/en)

Blue tilapia (*Oreochromis aureus*): Blue tilapia is native to North and West Africa but has also been introduced to other parts of the world for aquaculture and recreational fishing. Its bluish colouration recognizes it and is often found in freshwater lakes and rivers.

Picture 7: Blue tilapia
(https://www.inaturalist.org/taxa/121302-Oreochromis-aureus)

Mozambique tilapia (*Oreochromis mossambicus*): This species is native to southeastern Africa but has been introduced to various countries for aquaculture and biological control of aquatic weeds. Mozambique tilapia is known for its ability to adapt to brackish and even hypersaline waters.

Picture 8: Mozambique tilapia (https://mexican-fish.com/mozambique-tilapia/)

Red tilapia: Red tilapia is not a specific species but rather a hybrid resulting from crossing different tilapia species. These hybrids often have reddish colouration and are bred for their desirable traits, such as rapid growth and disease resistance.

Picture 9: Red tilapia (https://www.istockphoto.com)

Culinary Use: Tilapia is famous for its mild flavour and tender white flesh. It is versatile in cooking and can be prepared in various ways, including grilling, baking, frying, and steaming. Due to its neutral taste, tilapia readily takes on the flavours of different seasonings and marinades.

Aquaculture: Tilapia is one of the most extensively farmed fish globally, especially in countries with limited access to coastal marine resources. They are raised in aquaculture systems such as ponds, tanks, and cages. Tilapia farming is relatively efficient and can be practised in traditional and modern large-scale farms.

Health Benefits: Tilapia is a good source of lean protein, low in saturated fat, and contains essential nutrients such as vitamins and minerals. It is considered a healthy food choice in various balanced diets worldwide.

Pangasius

Pangasius, also known as pangas, basa, or swai, is a type of freshwater fish belonging to the Pangasiidae family. The most commonly farmed and marketed species of pangasius is *Pangasianodon hypophthalmus*. Native to Southeast Asia, particularly the Mekong River basin, pangasius has gained popularity in the global seafood market due to its relatively low cost, mild taste, and versatility in cooking (Rathod *et al.*, 2018). Pangasius has a slender, elongated body, flattened head, and slightly protruding lower jaw. The fish's colour can vary from grey to light pinkish-white, and it often has a white belly. Pangasius is famous for its mild flavour and firm, white flesh. It is a versatile fish that can be prepared in various ways, including grilling, baking, frying, and steaming. Due to its neutral taste, pangasius readily absorbs different seasonings and sauces, making it a preferred choice for various cuisines.

Picture 10: Pangasius
(https://www.ocean-treasure.com)

Aquaculture: Pangasius has become one of the most significant farmed fish globally, especially in countries like Vietnam and Thailand. Its popularity in aquaculture is primarily due to its fast growth rate, high feed conversion efficiency, and ability to tolerate various environmental conditions. Pangasius farming is commonly carried out in large-scale, intensive systems such as ponds and cages.

Asian sea bass (*Lates calcarifer*)

Asian sea bass, also known as barramundi or Asian barramundi, is a popular and commercially important fish species in the Indo-Pacific region. Scientifically, it is known as *Lates calcarifer* and belongs to the family Latidae. Barramundi is highly valued for its mild, white flesh and versatility in cooking, making it a sought-after fish in the culinary world. (Harpaz *et al.*, 2003). Asian sea bass has an elongated, muscular body with a slightly concave head and a prominent lower jaw. It has a silver to greyish colouration with dark vertical bars along its sides, which fade as the fish grows.

Habitat: Barramundi is typically found in brackish and freshwater habitats, including estuaries, mangroves, rivers, and lakes. While they spend much of their early life in estuaries and river mouths, adult barramundi often move upstream into freshwater rivers and downstream into the sea.

Diet: Barramundi are opportunistic predators and have a diverse diet that includes small fish, crustaceans, molluscs, and other aquatic creatures. Their feeding habits change as they grow, with more prominent individuals primarily preying on fish.

Fishing and Aquaculture: Barramundi is caught in the wild and commercially farmed. Due to its size and fighting abilities, it is a popular target for recreational anglers in the wild. On the commercial side, barramundi is extensively farmed in aquaculture operations in countries like Australia, Thailand, Vietnam, and Indonesia. The farming of barramundi has increased over the years due to its popularity as a food fish and its ability to adapt to aquaculture environments.

Culinary Use: Barramundi is highly regarded for its sweet, mild flavour and moist, flaky texture. It is a versatile fish that can be prepared in various ways, including grilling, baking, frying, steaming, and even in raw dishes such as ceviche. Barramundi's neutral taste allows it to take on the flavours of different seasonings and sauces, making it a favourite in various cuisines.

Health and Nutritional Considerations: Pangasius is a good source of lean protein and contains essential nutrients like vitamins and minerals. However, there have been concerns about the environmental impacts and the use of chemicals and antibiotics in some pangasius farming practices.

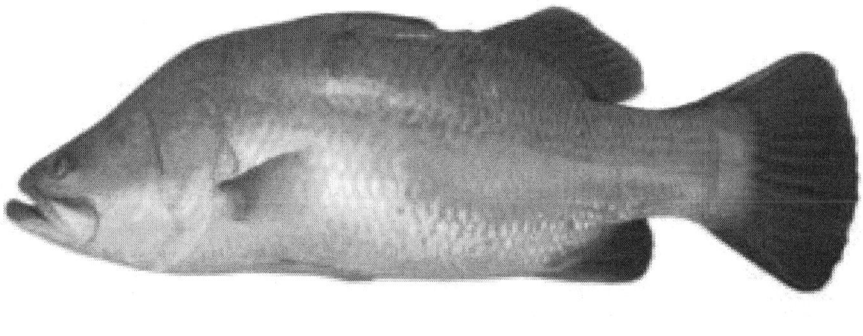

Picture 11: Asian Sea bass (https://mpeda.gov.in/?page_id=746)

Culturable Fish Species Worldwide:

Nile tilapia (*Oreochromis niloticus*)

Nile tilapia (*Oreochromis niloticus*) is one of the world's most essential and widely cultured fish species, particularly in aquaculture. It is native to Africa and has been introduced to many other regions due to its popularity in fish farming (Azevedo-Santos *et al.*, 2011).

Aquaculture suitability: Nile tilapia is well-suited for aquaculture due to its fast growth rate, adaptability to various environmental conditions, and high reproductive capacity.

Warm water fish: Nile tilapia is a tropical fish species that thrives in warm water temperatures ranging from 25°C to 35°C (77°F to 95°F). It can tolerate various salinities, making it suitable for fresh and brackish aquaculture.

Pond culture: Nile tilapia can be raised in ponds, which is the most common and traditional culture system. Ponds should be well-managed with proper aeration, water quality monitoring, and feeding regimes to ensure optimal growth.

Cage culture: This method is also used in areas with limited land or water resources. This method raises fish in floating cages or net pens in natural or artificial water bodies.

Feeding: Tilapia are omnivorous, meaning they eat plant and animal matter. In aquaculture, they are often fed with commercially prepared feeds, which may include fish meal, soybean meal, grains, and other nutritional ingredients.

Breeding and fry production: Tilapia can be bred through natural spawning or induced breeding. The eggs are incubated until they hatch into fry, which are then reared in nursery ponds or tanks before being transferred to grow-out ponds.

Harvesting: Nile tilapia can reach marketable size in about 6 to 9 months, depending on factors like water temperature, feeding practices, and stocking density. Harvesting can be done by draining the pond or using nets to catch the fish.

Market demand: Nile tilapia is in high demand globally due to its mild flavor and firm texture. It can be consumed fresh, frozen, or processed into various products like fillets, fish nuggets, and fish balls.

Rainbow trout (*Oncorhynchus mykiss*)

Rainbow trout (*Oncorhynchus mykiss*) is a popular and widely known freshwater fish species that is native to the Pacific coast of North America. Here are some critical points about rainbow trout (Hardy, 2002).

Habitat: Rainbow trout are primarily found in cold, clear streams, rivers, and lakes with good water quality. They prefer water temperatures between 10°C to 20°C (50°F to 68°F) but can tolerate a wide range of conditions.

Physical characteristics: *Rainbow trout* have torpedo-shaped bodies with small black spots on their silver sides. Their back can have shades of green to blue, and they have a pinkish-red stripe along their lateral line, which gives them their name "rainbow." (Behnke *et al.*, 2002).

Diet: Rainbow trout are opportunistic feeders and eat various foods, including insects, crustaceans, small fish, and zooplankton.

Sport fish: Rainbow trout are highly prized as sport fish due to their aggressive nature, strong fighting abilities, and willingness to take artificial lures and flies. They are stocked in many freshwater bodies worldwide to support recreational fishing.

Aquaculture: Rainbow trout have also been extensively raised in aquaculture for commercial purposes. They are farmed in freshwater ponds, raceways, and recirculating systems to supply the demand for their tasty, nutritious flesh.

Life cycle: Rainbow trout follow an anadromous life cycle, which means they are born in freshwater streams, migrate to the ocean to grow and mature, and return to their natal streams to spawn. However, not all rainbow trout populations exhibit this behaviour, and some remain entirely in freshwater throughout their lives (Ridolfi & Roles, 2006).

Conservation concerns: In some areas, rainbow trout introductions have adversely affected native fish populations. Invasive rainbow trout can compete with or prey upon native fish species, leading to declines in local biodiversity.

Fillet and culinary use: Rainbow trout has a delicate, mild flavour and a tender, flaky texture. It is a popular choice for grilling, baking, or pan-frying. The flesh can range from white to pink, depending on their diet and environment.

Picture 12: Rainbow trout
(https://www.flyfisherman.com/editorial/all-about-rainbow-trout/454217)

Atlantic salmon (*Salmo salar*)

Atlantic salmon (*Salmo salar*) is a species of migratory fish native to the rivers of the North atlantic ocean. It is a highly valued and sought-after fish for commercial and recreational purposes. Here are some critical points about Atlantic salmon (Stabell, 1984).

Habitat: Atlantic salmon are anadromous fish, meaning they are born in freshwater rivers and streams, spend a portion of their lives in the ocean, and return to their natal rivers to spawn. They are found in cold, clear, and well-oxygenated waters. (Johnson & Chalupnicki , 2014).

Life cycle: Atlantic salmon follow a complex life cycle. Adult salmon migrate upstream to their natal rivers to spawn, often overcoming significant obstacles like waterfalls and rapids. After spawning, the adult fish, known as "kelts," return to the ocean. The eggs hatch into alevins, which remain in the riverbed until they grow into fry and smolts. Smolts then go downstream to the ocean, where they spend a few years maturing before returning to their natal rivers as adults.

Conservation status: Wild Atlantic salmon populations have significantly declined due to habitat degradation, overfishing, and other factors. As a result, many populations are listed as endangered or threatened. Conservation efforts focus on habitat restoration, sustainable fishing practices, and hatchery programs to support population recovery.

Commercial fishing: Atlantic salmon is one of the most valuable fish species in the world. Commercial salmon fishing occurs in the ocean during the adult phase of their life cycle. Salmon are caught using nets or by trolling.

Aquaculture: Due to the high demand for salmon, especially in fillets, much of the Atlantic salmon consumed today comes from aquaculture. Farmed salmon are raised in net pens in coastal waters and fed a diet of fish meal, fish oil, and other protein sources.

Culinary use: Atlantic salmon is known for its rich, flavorful flesh, which ranges from pink to orange. It is famous for grilling, baking, smoking, and sushi. The high-fat content gives it a tender and moist texture.

Environmental concerns: The expansion of salmon farming has raised concerns about its environmental impact. Escaped farmed salmon can interbreed with wild salmon, potentially weakening the genetic diversity of wild populations. Additionally, the discharge of waste and chemicals from salmon farms can affect the marine ecosystem.

Recreational fishing: Atlantic salmon is a prized fish for recreational anglers, particularly for fly fishing. Many anglers participate in the catch-and-release fishery to support conservation efforts and ensure the sustainability of wild salmon populations.

Picture 13: Atlantic Salmon (*Salmo salar*) (https://goodfishbadfish.com/)

Japanese amberjack (*Seriola quinqueradiata*)

Japanese amberjack is a species of fish belonging to the Carangidae family. It is also commonly referred to as "yellowtail" or "hamachi" in Japan (Higuchi *et al.*, 2017). This fish is highly valued for its delicious taste and is a popular ingredient in many Japanese dishes, particularly sushi and sashimi.

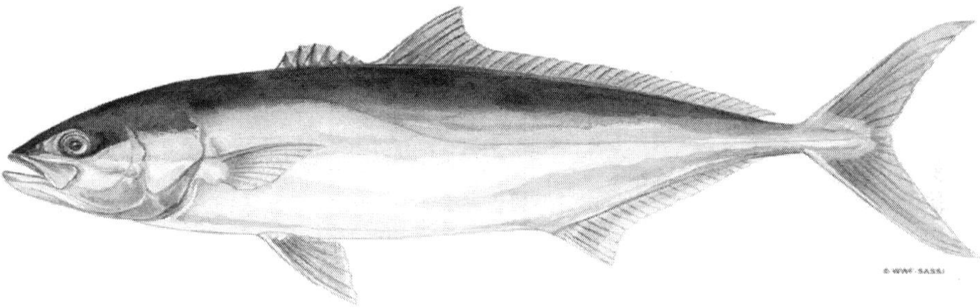

Picture 14: Japanese amberjack (https://seafood-guide)

Appearance: Japanese amberjacks have streamlined, elongated bodies. Their upper bodies are a distinct yellowish-to-golden colour, and their belly is a lighter shade. The fins and tail are usually yellow, contributing to their common name, "yellowtail."

Habitat: This species is native to the western Pacific Ocean and can be found in the coastal waters of Japan, Korea, China, and other neighbouring countries. It inhabits inshore and offshore waters and is often found near rocky reefs, kelp beds, and artificial structures.

Diet: Japanese amberjacks are carnivorous and feed on various prey, including smaller fish, squid, crustaceans and plankton.

Fishing and Aquaculture: Due to its popularity as a food fish, Japanese amberjack is extensively caught through commercial and recreational fishing. Additionally, it is one of the most important fish species in aquaculture in Japan, where it is raised in fish farms to meet the high demand for its consumption.

Culinary use: Japanese amberjack is prized for its tender and fatty flesh, which has a rich, buttery flavour. It is often served raw as sashimi, where it is thinly sliced and accompanied by soy sauce, wasabi, and pickled ginger. It is also commonly found in sushi rolls and grilled preparations, where various seasonings enhance its natural flavours.

Health benefits: Amberjack is a good source of high-quality protein, omega-3 fatty acids, vitamins (such as B-complex vitamins), and minerals (including selenium and potassium). Like many fish, it is considered a healthy addition to a balanced diet.

Channel catfish, scientifically known as *Ictalurus punctatus*, is a freshwater catfish native to North America. It is one of the most popular and widely distributed catfish species in the United States and is highly valued for recreational fishing and aquaculture purposes.

Appearance: Channel catfish have a distinctive appearance, characterized by their slender, cylindrical bodies and deeply forked tail fins. Their colouration can vary but is typically olive-brown to bluish-grey on the upper body, with a lighter colour on the sides and belly. They have sharp spines on their dorsal and pectoral fins, which can cause painful stings if mishandled.

Habitat: Channel catfish are found in various aquatic habitats, including rivers, streams, lakes, reservoirs, and ponds. They prefer slow-moving or still waters with ample cover, such as submerged logs, vegetation, and rock structures.

Diet: Channel catfish are opportunistic omnivores with a diverse diet. They feed on various aquatic organisms, including insects, crustaceans, small fish, molluscs, and plant matter.

Fishing: Anglers seek channel catfish due to their large size, strong fighting abilities, and excellent table fare. Anglers often use a variety of baits, including worms, cut bait, chicken liver, and stink baits, to catch these fish. They are typically caught using traditional fishing methods such as rod and reel.

Aquaculture: Channel catfish are also an essential species in aquaculture. They are widely raised in fish farms throughout the United States to meet the demand for their meat in the food market. Bonvechio *et al.*, 2016). Catfish farming has become a significant industry, and the fish are typically grown in large ponds or tanks.

Culinary use: Channel catfish is renowned for its mild, white, and flaky flesh. It is a popular food fish and is commonly prepared in various ways, including fried, grilled, baked, or sautéed. In the Southern United States, catfish is a staple dish in many traditional recipes. (Nakada, 2008)

Conservation: While channel catfish populations remain relatively stable due to their popularity in aquaculture, it is still essential to manage wild populations sustainably. Overfishing, habitat degradation, and water pollution can_impact their populations in the wild, so conservation efforts are necessary to ensure their continued existence in their natural habitats.

References

Ahmed., N, Allison E.H. & Muir., J. F. (2010) Rice-fields to prawn farms a blue revolution in southwest Bangladesh Aquacult Int, 18:555–574

Azevedo-Santos, V. M. D., Rigolin-Sá, O. & Pelicice, F. M. (2011). Is it growing, losing or introducing? Cage aquaculture as a vector for the introduction of non-native fish in Furnas Reservoir, Minas Gerais, Brazil. *Neotropical Ichthyology*, *9*, 915-919.

Behnke, Robert J. (2002). "Rainbow and Redband Trout". *Trout & Salmon of North America*. Tomelleri, Joseph R. (illustrator). New York: The Free Press. pp. 65–122. ISBN 978-0-7432-2220-4.

Béné C, Arthur R, Norbury H, Allison EH, Beveridge M, Bush S, Campling L, Leschen W, Little D, Squires D, Thilsted SH. & Troell M, Williams M (2016) Contribution of fsheries and aquaculture to food security and poverty reduction: assessing the current evidence. *World Dev, 79*, 177–196

Bonvechio, T. F., Marsik, J. E. & Bussells, C. W. (2016). Population dynamics of introduced Flathead Catfish in two Atlantic coastal plain rivers under differing management strategies. *Journal of the Southeastern Association of Fish and Wildlife Agencies, 3*, 128-135.

Bovenkerk, B., & Meijboom, F. L. (2012). The moral status of fish. The importance and limitations of a fundamental discussion for practical ethical questions in fish farming. *Journal of Agricultural and Environmental Ethics, 25*(6), 843-860.

Eknath, A. E. & Hulata, G. (2009). Use and exchange of genetic resources of Nile tilapia (Oreochromis niloticus). *Reviews in Aquaculture, 1*(3 4), 197-213.

FAO (2020) The state of world fsheries and aquaculture: sustainability in action. Food and Agriculture Organization of the United Nations, Rome.

García-Berthou, E. (2001). Size-and depth-dependent variation in habitat and diet of the common carp (*Cyprinus carpio*). *Aquatic Sciences, 63*, 466-476.

Ghosh, R., Upadhyay, A. D., Roy, A. K., & Tiwari, A. (2020). Structural and functional analysis of cytochrome b protein of Indian major carps *Labeo rohita*. *Journal of Entomology and Zoology Studies, 8*(2), 540-550.

Hardy, R. W. (2002). Rainbow trout, *Oncorhynchus mykiss*. In *Nutrient requirements and feeding of finfish for aquaculture* (pp. 184-202). Wallingford UK: CABI Publishing

Harpaz, S., Glatman, L., Drabkin, V. & Gelman, A. (2003). Effects of herbal essential oils used to extend the shelf life of freshwater-reared Asian sea bass fish (*Lates calcarifer*). *Journal of Food Protection*, 66(3), 410-417.

Higuchi, K., Yoshida, K., Gen, K., Nyuji, M., Takashi, T., Mushiake, K. & Soyano, K. (2017). Effect of timing of restricted feeding on sexual maturation in female yellowtail, *Seriola quinqueradiata*. *Aquaculture, 479*, 609-615.

Hora S.L., Pillay T. (1962). Hand book on fish culture in the Indo-Pacific region, 14-204.

Johnson, J. H., & Chalupnicki, M. A. (2014). Interspecific habitat associations of juvenile salmonids in Lake Ontario tributaries: implications for Atlantic salmon restoration. *Journal of Applied Ichthyology, 30*(5), 853-861.

Khan, H. A., & Jhingran, V. G. (1979). Synopsis of biological data on the mrigal *Cirrhinus mrigala* (Ham.). FAO Fish Synop, 120-78.

Nakada, M. (2008). Capture-based aquaculture of yellowtail. *Capture-based aquaculture. Global overview. FAO Fisheries Technical Paper, 508*, 199-215.

Neori A., Troell M., Chopin T., Yarish C., Critchley A. & Buschmann A. H. (2007). The need for a balanced ecosystem approach to blue revolution aquaculture. Env Sci *Policy Sustain Dev* 49(3):36–43

Rathod, N. B., Pagarkar, A. U., Pujari, K. H., Shingare, P. E., Satam, S. B., Phadke, G. G., & Gaikwad, B. V. (2018). Status of valuable components from Pangasius: a review. *International Journal of Current Microbiology and Applied Sciences*, *7*(4), 2106-2120.

Ridolfi, K. & Roles, E. (2006). Animal diversity web. https://animaldiversity.org/

Simpson S. (2011). The blue food revolution: making aquaculture a sustainable food source. *Scientific American*, 54–61

STABELL, O. B. (1984). Homing and olfaction in salmonids: a critical review with special reference to the Atlantic salmon. *Biological Reviews*, *59*(3), 333-388.

Stepien, C. A., Snyder, M. R. & Elz, A. E. (2019). Invasion genetics of the silver carp *Hypophthalmichthys molitrix* across North America: Differentiation of fronts, introgression, and eDNA meta barcode detection. PLoS One, 14(3), e0203012.

Yaqoob, S. (2021). A Review of Structure, Origin, Purpose & Impact of Common Carp (*Cyprinus carpio*) in India. *Annals of the Romanian Society for Cell Biology*, *25*(6), 34-47.

CHAPTER 6
INTEGRATED FISH FARMING

Introduction

In rural areas, land holdings are often small and scattered. Conventional large-scale production methods, which demand significant inputs, do not effectively address the issues of low income and productivity in these communities. Smallholders and marginal farmers typically possess livestock such as cattle, pigs, ducks, chicks, agricultural land and surplus family labour. To overcome these obstacles, cost-effective farming systems are devised to enhance productivity, optimize farm waste utilization, leverage available resources, and maximize human resources. Integrated farming systems have surfaced as a promising approach, intertwining fish culture, livestock rearing, and agricultural practices. Extensive research has led to developing and verifying economically viable and feasible practices for integrated farming at the farmer's level (Mutua, 2018).

Cultivating fish in fields where paddy, wheat, and coconut are grown is possible. Bearing. On the dykes, there is cultivation of both flowering plants and vegetable plants. Azolla: The practice of cultivating fish is gaining popularity as well.

Paddy cum- fish- culture

Paddy-cum-fish culture, a potentially lucrative endeavour, can generate significant profits for farmers with the proper management strategies. This method flourishes in paddy fields with ample water supply from rivers or lakes. In India, regions such as Kerala and West Bengal, located along the coast, have a rich history of engaging in paddy-cum-fish culture. Nevertheless, despite its considerable promise, this combined method of cultivating fish alongside freshwater paddy has not garnered widespread adoption in the country.

Despite India allocating six million hectares of land for rice farming, only a minuscule 0.03 per cent of this space is currently employed for rice-fish cultivation.

The primary cause for this limited uptake can be traced to the transition in farming methodologies, moving away from conventional practices towards modern approaches featuring high-yield rice strains and heightened pesticide usage. Additionally, the introduction of multiple cropping systems has further enhanced the productivity of agricultural lands, diverting the focus away from integrated farming practices like paddy-cum-fish culture (Haobijam *et al.*, 2018).

The integrated culture requires ample water and is best suited for low-lying areas. A broad expanse of water spreading across millions of hectares is highly advantageous for this comprehensive farming method. This approach facilitates the annual growth of two batches of rice and one batch of fish. Submerged rice fields provide an ideal natural environment for various types of fish. The presence of fish in the rice fields results in a boosted rice output ranging from 5 to 15 per cent. Fish consume significant amounts of unwanted vegetation, worms, insects, larvae, and algae, which can directly or indirectly damage rice crops. Furthermore, fish contribute to the availability of fertilizing materials, making them more easily absorbed by rice plants (Yani & Sharma 2018).

The advantage of paddy-cum-culture

- The economical utilization of land refers to the efficient and cost-effective use of land resources.
- Implementing this method requires minimal additional labour compared to traditional farming practices.
- By adopting this approach, there are savings on labour costs associated with activities like weeding and providing additional feed for the crops.
- One of the benefits is an enhanced rice yield, ranging from 5 to 15%. This increase is attributed to the indirect organic fertilization provided by fish excreta.
- The practice allows for the production of fish alongside the cultivation of paddy fields.
- Participating in this approach presents a chance to augment one's income and procure a varied yield, encompassing fish and rice sourced from water bodies, along with onion, bean, and sweet potato grown on elevated embankments known as bunds.
- Fish serve a vital function in managing undesirable filamentous algae, which, if left unchecked, would vie for nutrients with the cultivated crops.
- Including tilapia and common carp contributes significantly to suppressing unwanted aquatic weeds, thereby averting potential reductions in rice yields of up to 50%.
- Fish, particularly murrels and catfishes, effectively control insect pests of rice, such as stem borers, by feeding on them.

- Fish also contribute to the control of water-borne diseases by consuming aquatic intermediate hosts like malaria-causing mosquito larvae.
- Rice fields have the potential to function as breeding grounds for fish, enabling young fish to develop into larger juveniles. When there is a substantial output of these juveniles, they can be marketed or introduced into aquaculture ponds to enhance fish production, particularly when utilizing methods such as composite fish culture (Rai, 2005).

Site selection

The ideal amount of rainfall for this integrated system is approximately 80 cm. It is preferable to have fields with a consistently even contour and a high capacity for retaining water. When selecting a site, it is crucial to consider the groundwater table and the efficiency of the drainage system.

Types of paddy filed for the integrated farming system

1. In the Perimeter type, the paddy cultivation zone sits nestled within the heart of the integrated system. Positioned at a moderate elevation, the ground gently slopes in all directions, guiding excess water into perimeter trenches to ensure adequate drainage.
2. Conversely, in the Central Pond type, the paddy fields find their place along the outer periphery of the integrated system. Here, the land slopes inward, directing water flow towards a central pond as a focal point within the landscape.
3. Lateral trench type: Trenches are created along one or both sides of the paddy field with a moderate slope.

Considering that the total area of the integrated system is 100m x 100m (1 hectare), the area designated for paddy cultivation should measure 82m x 82m, amounting to 0.67 hectares. The area reserved for fish culture is 6m x 352m, equal to 0.21 hectares (on all four sides). The embankment area should have 3m x 388m dimensions, accounting for 0.12 hectares. Lastly, the area allocated for fruit plants should measure 1m x 388m, equivalent to 0.04 hectares. These proportions represent the recommended ratios for establishing an integrated system (Nayak *et al.*, 2018).

Paddy cultivation

Rice varieties used for the integrated system:

The most promising deepwater varieties chosen for different states are as follows:

1. PLA-2 (Andhra Pradesh)
2. AR-1, 353-146 (Assam)
3. BR-14 Jisurya (Punjab)

4. AR 61-25B, PTB-16 (Kerala)
5. TNR- 1, TNR (Tamil Nadu)
6. Jalamagan (Uttar Pradesh)

In areas where fish are cultivated alongside rice, farmers utilize Manoharsali rice seeds. The preparation of the paddy field typically commences in April-May. After field preparation, farmers opt for deepwater paddy varieties for direct sowing in low-lying regions immediately after the initial monsoon rainfall (Hossain *et al.*, 2005).

Fertilization schedule

In the cultivation of paddy, the paddy plots are supplemented with farm yard manure or compost at a rate of 30 tons per hectare as a basal dose. This enrichment helps to enhance the fertility of the soil. Since deep water paddy has a significant nutrient demand, applying inorganic fertilizers at specific rates is advisable. The recommended inorganic fertilizers for deep water paddy are nitrogen and potassium, both at a rate of 60 kilograms per hectare.

It is recommended that nitrogen and phosphorus be applied in three phases. These phases include planting, tilling, and flowering initiation. This staggered application ensures that the paddy plants receive an adequate and timely supply of these essential nutrients at different stages of their growth. By providing nitrogen and phosphorus in these key phases, paddy plants can effectively utilize the nutrients and support their overall development and productivity (Ding *et al.*, 2018).

Pesticide use

The development of Paddy-cum-fish culture has been hindered by the widespread use of pesticides in rice fields to eliminate various pests. Unfortunately, these pesticides are highly toxic to fish, posing a significant challenge. However, there are potential solutions to this pesticide problem, such as implementing an integrated pest control system and utilizing less harmful pesticides to fish, but only when necessary. Specifically, the use of carbamates and selective organophosphates is recommended. It has been observed that Furadon when administered seven days before fish stocking, has been proven safe for fish. These measures aim to balance pest control in rice fields and preserve fish health in Paddy-cum-fish culture.

During the Kharif crop period, it is advisable to refrain from using pesticides. The harvesting of the Kharif crop typically occurs between November and December. The expected yield for this crop ranges from 800 to 1200 kg per hectare.

Pesticides are allowed for Rabi crops—the need. Before applying pesticides on the paddy, the trench's embankment should be examined. The amount should be raised to prevent the pesticide from entering the trenches—the rice crop yields 4000 to 5000 kilograms per hectare (Berg & Tam, 2012).

Culturable species of fish in rice fields

Cultivating fish in rice fields necessitates species that can thrive in particular environmental settings, including shallow water depths (>15 cm), high temperatures (up to 35°C), low levels of dissolved oxygen, and high turbidity. Various fish species have proven successful in this environment, such as *Labeo rohita, Catla catla.* Composite and integrated fish farming, *Oreochromis mossambicus, Anabas testudineus, Clarias batrachus, Clarias macrocephalus, Channa striatus, Channa punctatus, Channa marulius, Heteropneustes fossilis, Chanos chanos, Lates calcarifer*, and *Mugil sp.* These species have demonstrated their ability to adapt to the conditions commonly found in rice fields. Smaller carp species like *Labeo bata, Labeo calbasu, Puntius japanicus, Puntius sarana*, etc., can also thrive in paddy fields. Another viable option for cultivation in rice fields is the freshwater prawn *Macrobrachium rosenbergii.* The appropriate species for cultivation depends on water depth, duration, and the paddy varieties' specific characteristics (Halwart & Gupta, 2004).

The major system of paddy -cum fish culture

Two major systems of paddy -cum culture that can be undertaken in freshwater areas are:

1. Paddy- cum- carp culture:
2. Paddy-cum-air-breathing fish culture:

1. Paddy- cum- carp culture

In paddy fields, both significant and minor carp are cultivated. During July, when rainwater begins to accumulate in the paddy fields, and the water level in the waterway reaches an adequate depth, the fish are introduced into the ecosystem. The stocking density ranges from 4000 to 6000 fish per hectare. Maintaining a specific species ratio is recommended to ensure a balanced fish population. Approximately 25% of the stocked fish should be surface feeders, preferably catla. Another 30% should be column feeders, such as rohu, while the remaining 45% should consist of bottom feeders like mrigal or common carp. This distribution helps create a diverse aquatic community within the paddy fields (Ghosh *et al.*, 1984).

2. Paddy -cum- air-breathing fish culture

Air-breathing catfish, such as Singhi and Magur, flourish in the tending of paddy plots discovered in many rice-raising areas. The inundated milieu established in these plots yields optimal circumstances for the swift development of these varieties. In this unified structure, an equivalent count of Magur and Singhi juveniles should be introduced at a concentration of one fish per square meter. Channa types are also apt for this manner of farming (Goswami *et al.*, 2004).

Picture1: Paddy-cum-fish culture
(Source: https://www.agrifarming.in/rice-fish-farming)

Fish culture in rice fields methods

1. Simultaneous culture

Simultaneous culture involves co-cultivation of rice and fish in the same rice plots. Typically, a 0.1-hectare area can efficiently accommodate this practice. Four rice plots are usually established Within this space, each measuring 250 square meters (25 by 10 meters). These plots have ditches measuring 0.75 meters in width and 0.5 meters in depth. Surrounding the rice plots are dykes, which are typically 0.3 meters high and wide, fortified by embedding straw within them to enhance their strength. Besides serving as a repository for fish waste when the fish aren't foraging amongst the rice plants, the ditches also serve as collection channels, where fish gather during water level fluctuations. The water depth within the rice plot varies between 5 and 25 centimetres, depending on the specific rice variety and the fish species being cultivated.

Five days after transplanting the rice, fish fry is introduced into the area at a density of 5000 per hectare, or fingerlings can be stocked at a rate of 2000 per hectare. If supplemented with daily feed, the stocking density can be doubled. Paddy-cum-fish culture, integrating rice and fish cultivation simultaneously, presents numerous advantages, detailed under the heading 'Advantages of Paddy-Cum-Fish Culture.' However, constraints are also associated with the concurrent cultivation of fish and rice, as discussed by (Nair *et al.*, 2010).

1. Using agrochemicals is often not feasible due to potential harm to the fish.
2. Maintaining a high water level may not always be possible, especially when considering the size and growth requirements of the fish.
3. Fish species like grass carp may feed on rice seedlings, posing a challenge to their growth.

4. Fish species like common carp and tilapia may uproot the rice seedlings, further impacting rice cultivation.

2. Rotational culture of rice and fish

In this specific system, fish and rice are grown alternately. After the rice harvest, the rice field is converted into a temporary fish pond. This sequential method is favoured over simultaneous cultivation because it allows the application of pesticides and herbicides to rice cultivation. Furthermore, it enables the maintenance of a deeper water level, up to 60 cm, throughout the fish farming period.

Typically, the field is prepared for fish cultivation one to two weeks after rice harvest. The recommended stocking densities for this method are 20,000 fry per hectare or 6,000 fingerlings per hectare, depending on the chosen approach (composite or integrated fish farming). It is worth noting that the fish yield in this rotational culture system can potentially exceed the income generated from rice cultivation (Halwart & Gupta 2004).

Fish culture

To maintain the health and productivity of the crops, a systematic approach is taken to tackle weeds and predators in trenches and paddy fields. The weeds are carefully removed by hand, ensuring that they are thoroughly eliminated from the area. Similarly, predatory and weed fishes are also targeted for removal. This can be achieved through two methods: netting or dewatering. These unwanted fishes can be effectively eliminated by using nets or draining the water. Mohua oil cake is recommended to combat predatory and weed fishes further. This cake should be applied at a concentration of 250 ppm (parts per million). This treatment helps eradicate the presence of predatory and weed fishes, contributing to a healthier environment for the crops (Sastry *et al.*, 2014).

Once the weeds and predators have been successfully dealt with, providing the crops with appropriate fertilizers is essential. The fertilizers should be applied in equal portions throughout the rearing period. The recommended fertilizers and their respective quantities are as follows:

1. Cow dung: 5000kg/ha
2. Ammonium sulphate: 70kg/ha
3. Single superphosphate: 50kg/ha

In simultaneous and rotational culture practices, stocking density varies, as mentioned in the sections above. The fish are supplemented with food in different ratios depending on the culture system. In paddy-cum-carp culture, the supplementary food provided to the fish consists of a mixture of rice bran and groundnut oil cake in a 1:1 ratio. The amount of food given is 5% of the body weight of the fish. In paddy-cum-air breathing culture, the fish are provided with a mixture of fish meal and rice bran in a ratio of 1:2. The feeding rate is also 5% of the body weight of the fish (Bajwa *et al.*, 2015).

Once the paddy is harvested and the plots gradually dry up, the fish seek shelter in the waterway. Partial harvesting begins shortly after the Kharif season using drag netting. Fish that have reached their maximum size are removed every two weeks. When the water in the waterway irrigates the Rabi paddy, the remaining fish are manually selected. The fish production in this combined system varies from 700 to 1000 kg per hectare.

Nonetheless, the survival rate of the fish is below 60%. It's noted that the survival rate is highest in rejuvenated paddy plots compared to fish cultivation in regular paddy plots. The dams built for this system can be utilized to cultivate vegetables and fruit-bearing plants such as papaya and bananas to maximize returns. Additionally, fish can be raised alongside wheat, a practice observed in Madhya Pradesh. Similarly to paddy fields, the same fish species can be cultivated in wheat fields using management techniques akin to fish-cum-paddy culture. Fish can also be farmed alongside coconut trees.

Fish cum horticulture

A sizable portion of an aquaculture farm can be found as dykes, some of which are used for typical farm tasks and the rest of which is left fallow year-round and infested with deep-rooted terrestrial organisms. In addition to hindering regular farm operations, the menacing growth of these weeds necessitates ongoing weed control expenses. This harms aquafarming's economy, which could be significantly improved by using dykes wisely to produce vegetables and fish feed. Farming strategy that integrates horticulture and aquaculture results in better resource management.

Cabbage, cauliflower, tomato, brinjal, coriander, turnip, radish, beans, spinach, fenugreek, bottle gourd, potato, and onion are examples of winter vegetables. Some summer vegetable varieties include amaranth, water bindweed, papaya, okra, bitter gourd, sponge gourd, sweet gourd, and ridge gourd. Depending on the embankments' dimensions, shape, and condition, certain crops such as gourd, chillies, ginger, and turmeric can be cultivated (Gupta *et al.*, 2015).

Carp production using leafy vegetables and vegetable wastes

Many cabbage, cauliflower, turnip, and radish leaves are typically discarded during harvest. However, these leafy vegetables can be effectively utilized as supplementary feed for grass carp, resulting in profitable outcomes. In winter, turnip, cabbage, and cauliflower leaves can be used as feed for grass carp. At the same time, during summer, amaranth and water-bind weed can be provided as supplementary feed through fortnightly clipping, aiding in the rearing of grass carp. By adopting a monoculture approach of grass carp, with a stocking density of 1000 fish per hectare, and solely feeding them vegetable leaves, an average production of approximately 2 tons per hectare per year can be achieved. On the other hand, practising mixed culture by combining grass carp with rohu, catla, and mrigal fish in a ratio of 50:15:20:15, respectively, at a higher density of 5000 fish per hectare, can result in an average production of 3 tons per hectare per year (Maucieri *et al.*, 2019).

Integrated farming, combining dairy, piggery, and poultry, has been widely practised globally, albeit with different outcomes. In India, including freshwater fish cultivation in such integrated systems has become necessary, primarily for recycling organic waste and fostering rural development. Beyond the traditional practice of using cattle farm waste as pond fertilizer, substantial amounts of waste from poultry, duckery, piggery, and sheep farming are easily accessible. These wastes, richer in nutrients than cattle waste, can significantly boost fish production even in smaller quantities.

Integrated fish – cum poultry farming

There's a notable emphasis on the expansion and progress of poultry farming in India. Through improved scientific management methods, poultry farming has become increasingly popular as a rural business in various states nationwide. Alongside the production of eggs and chicken, poultry farming also generates a valuable by-product of manure, which holds significant fertilizer value. The estimated output of poultry droppings in India is roughly 1,300 thousand tons, equivalent to about 390 metric tons of protein. Utilizing this abundant resource as manure in aquaculture has the potential to achieve better conversion rates compared to its application in traditional agricultural practices (Misra *et al.*, 2019).

Picture 2: Fish – cum poultry farming
(Source: https://www.google.com)

Pond management

Pond management encompasses a range of activities to preserve the well-being and productivity of a pond's ecosystem. A vital component of this management involves removing aquatic weeds, undesirable fish species, and insects.

Aquatic weeds are plant species that grow abundantly in water bodies, such as ponds. While some weeds may provide benefits like oxygen production and habitat for marine organisms, excessive growth can lead to problems. Overgrown aquatic weeds can impede water circulation, hinder recreational activities, and disrupt the balance of the ecosystem by consuming excessive nutrients.

Unwanted fish may not be desirable or compatible with the goals of the pond management plan. For example, certain invasive fish species can outcompete native fish species, causing imbalances in the ecosystem. Removing unwanted fish helps maintain the desired fish population and prevent ecological disruptions.

Insects, too, can affect the pond ecosystem. Some insects may harm the pond's health, such as mosquito larvae that can multiply rapidly and become a nuisance. Additionally, certain insects can prey on desirable aquatic organisms or disrupt the ecosystem's natural balance. Managing insect populations can help mitigate these issues and promote a healthier pond environment (Oben *et al.*, 2015).

Stocking

The use of poultry manure in the pond promotes dense phytoplankton growth, particularly nanoplankton, which supports the development of zooplankton. Zooplankton benefits from bacteria that thrive on the organic matter from the added poultry waste, providing them with an extra food source. This highlights the importance of stocking phytoplankton-eating and zooplankton-eating fish in the pond. Furthermore, a considerable quantity of organic matter gathers at the bottom of the pond, acting as a foundation for tiny organisms and bottom-dwelling creatures, particularly chironomid larvae. Consequently, it's crucial to focus on introducing bottom-feeding species.

Additionally, grass carp would make a beneficial addition to the pond as they consume aquatic plants. If macrophytes are lacking, they can be supplemented with green cattle fodder cultivated on the pond's banks. The partially digested excrement of these fish serves as food for the bottom feeders (Gomes, 2003).

Fish and poultry ponds employ a polyculture approach to efficiently utilize food resources, incorporating three major Indian carp and three exotic carp. Before stocking, the pond water undergoes detoxification. Stocking rates typically range from 8000 to 8500 fingerlings per hectare. A preferred species ratio is recommended for optimal fish yields, comprising 40% surface feeders, 20% column feeders, 30% bottom feeders, and 10-20% weedy feeders. Alternatively, a mixed culture featuring Indian major carp can be practised, with a species ratio of 40% surface feeders, 30% column feeders, and 30% bottom feeders.

In India's northern and north-western states, ponds are ideally stocked in March and harvested in October-November due to the adverse impact of severe winters on fish growth. Conversely, stocking should occur between June and September in regions such as the south, coastal areas, and northeastern states with milder winters. Following a rearing period of 12 months, harvesting can take place (Oben *et al.*, 2015).

Use of poultry litter as manure

Two methods are adopted in recycling poultry manure for fish farming.

1. Farm poultry waste is gathered, stored, and spread onto ponds daily following sunrise at 50 kilograms per hectare per day. Application is

postponed when there is an algal bloom to manage the utilization of the waste.

2. Poultry housing structures partially cover fish tanks and recycle droppings for fish culture. However, direct recycling and excessive manure can lead to oxygen depletion and fish mortality.

Every year, a quantity of one ton of deep litter fertilizer is generated by approximately 30 to 40 birds. To illustrate, a group of 500 birds weighing 450 kg can produce approximately 25 kilograms of wet manure daily, ample for nurturing a hectare of water area in polyculture. This deep litter is composed of 3% nitrogen, 2% phosphate, and 2% potash. It is readily accessible in extensive poultry operations. Farmers lacking poultry infrastructure can procure poultry litter for application in their fields (Debnath *et al.*, 2013).

Harvesting

In some fish farming practices, specific fish species reach a marketable size within a few months. This is considered by considering factors such as the fish size, prevailing market rates, and local demand for the fish. As a result, a technique known as partial harvesting is employed for table-size fish. Following a partial harvest, the pond restocks with an equivalent number of fingerlings of the same species, dependent on the fish seed availability. The ultimate harvest usually occurs after a 12-month rearing period.

The fish yield can vary depending on the stocking strategy. When stocking six species, the average fish yield ranges from 3500 to 4000 kilograms per hectare per year. On the other hand, when stocking only three species, the average fish yield ranges from 2000 to 2600 kilograms per hectare per year.

In poultry farming, eggs are gathered twice daily, once in the morning and again in the evening. On average, each bird produces approximately 200 eggs annually. Nonetheless, the birds are typically marketed for their meat once raised for 18 months. This timing is due to the gradual decline in egg-laying capability observed in these birds beyond this timeframe (Naugi *et al.*, 2007).

Integrated fish -cum- duck farming

Integrated fish-cum-duck farming, a method extensively employed in China, is currently garnering traction in India, notably in areas like West Bengal, Assam, Tamil Nadu, Andhra Pradesh, Kerala, and Bihar, among others. This approach involves combining fish farming with duck rearing, leveraging the advantages of their mutual biological relationship. Ducks utilize land and water habitats, and integrating them with fish maximizes the benefits of this symbiotic association. This approach supports the ducks' well-being and boosts fish farming by providing extra organic fertilizers. The combination of fish and ducks showcases the possibility of attaining notable economic effectiveness in fish farms.

Picture 3: Fish- cum- duck farming
(Source: Popp *et al.*, 2018)

Ducks feed on pond organisms like insect larvae, tadpoles, molluscs, and aquatic weeds. They don't eat the stocked fish. Duck droppings fertilize the pond, and their dabbling releases nutrients from the soil, boosting pond productivity. This results in a rise in fish yield. In 1934, German farmer Probst initiated the inaugural trials on combined fish and duck farming, paving the way for integrated systems. Ducks enjoy a clean environment and natural food from the pond, promoting their growth (Pandey *et al.*, 2016).

Benefits of fish cum -duck farming

- Duck raising allows for the full utilization of water surface in ponds.
- Fish ponds create a favourable environment for ducks, protecting them from parasite infections.
- Ducks consume predators and contribute to the growth of fingerlings.
- Raising ducks in fish ponds decreases the protein required in duck feed to 2-3%.
- Duck excrement contributes essential nutrients directly to the water, fostering the development of natural food sources. The leftover duck feed, typically 20-30 grams per duck per day, can serve as fish feed in ponds or be utilized as fertilizer, leading to increased fish production.
- Ducks distribute their manure evenly throughout the pond without forming piles.

- By digging in search of benthos (bottom-dwelling organisms), ducks diffuse soil nutrients into the water, stimulating plankton production.
- Ducks act as aerators by swimming, playing, and chasing each other, which facilitates the aeration of the pond's surface.
- Ducks exhibit improved feed efficiency and body weight; fish can consume spilt feeds.
- Ducks raised in fish ponds experience a 3.5% increase in survival rate due to the clean environment provided by the fish ponds.
- Combining duck droppings and leftover feed can increase fish output to 37.5 kilograms per hectare.
- Ducks play a vital role in managing the proliferation of aquatic vegetation within the pond.
- Duck farming doesn't demand extra land allocation.
- Integrating duck farming with fish ponds results in increased fish, duck eggs, and duck meat yields within a defined period and water surface.
- This approach ensures high profitability with reduced investment (Paul *et al.*, 2018).

Integrated fish-cum-pig farming

The practice of raising pigs with fish involves the construction of pig sties on or near the pond embankment. This allows for direct drainage of pig waste into the pond or for the waste to be manually applied. The pig dung serves as an effective pond fertilizer, enhancing the biological productivity of the pond and ultimately increasing the fish yield. Interestingly, the fish directly consume pig manure, a highly digestible feed comprising 70% of their diet. Consequently, no additional fish or pond fertilization is required within this integrated system.

Picture 4: Fish- cum- pig farming (Source: https://agritech.tnau.ac.in)

This approach significantly cuts down on the costs related to fish farming. Pig waste replaces fish feed and pond fertilizers, which usually make up 60% of the total expenses in traditional fish farming techniques. This combined method is crucial as it could enhance the socio-economic conditions of rural areas, predominantly tribal communities involved in pig farming traditionally (Avinla, 2004).

Benefits of fish cum -pig farming

- The fish thrive on the rich nutrients found in the leftover food and waste from pigs, which provide a valuable source of nutrition.
- By incorporating pig manure as an alternative to traditional pond fertilizers and supplementing fish feed, the expenses associated with fish farming are considerably diminished.
- Piggery operations do not require any additional land.
- The pond embankments are used to grow cattle fodder for pigs and grass.
- The pond serves as a water source for cleaning the pig sties and providing water for the pigs.
- This approach leads to a significant increase in the production of animal protein within a given space.
- It guarantees a substantial profit margin while requiring minimal initial investment.
- The sediment buildup in the ponds, caused by the ongoing use of pig waste, can be repurposed as fertilizer for cultivating vegetables, various crops, and fodder for livestock (Saha & Nath, 2013).

Integrated fish-cum – cattle farming

Fish farming using raw cow manure is one of the standard practices worldwide. Among all livestock manure, cow dung is most abundant throughout rural India. A healthy cow weighing 400-450 kg excretes over 400-500 kg of dung and 3,500-4,000 litres of urine annually due to its quick sinking ability. The nutrients available are responsible for increased natural food organisms, which are detrimental and beneficial bacteria in fishponds. The edible, non-digested feed of cow dung is consumed directly by fish. The biological oxygen demand (BOD) of cow manure is lower than that of other livestock manures as it is already decomposed by microorganisms in ruminants. The nutritional value of excreta of different animals is given in. The faeces and urine are highly beneficial for filter feeding and omnivorous fish such as catla and silver carp. A unit of 5-6 cows can provide adequate dung and urine to produce 3,000-4,000 kg of fish per ha per year. With these types of integration, 9,000 litres of milk and fish harvested from the system became most popular among rural households. The shed of cattle, built on the dykes or near the pond, helps in the easy disposal of urine and dung into the pond simplifying the handling problem. The fresh dung and urine collected separately can be applied periodically @ 10 tonnes/ ha/year to culture ponds (Debnath *et al.*, 2013).

Integrated fish-cum – prawn culture

Despite significant efforts invested in composite fish culture, which combines Indian major carp and exotic carp with distinct feeding habits and achieves considerable production, there has been a lack of large-scale attempts to polyculture prawns and fish. However, there is potential for implementing polyculture practices in Indian waters by cultivating surface and column-feeding carp together with bottom-feeding prawns, aiming to maximize yield. It is worth noting that the culture of carp and freshwater prawns is more prevalent in this polyculture system compared to the cultivation of brackish water prawns alongside other fish species (Mirhaj *et al.*, 2014).

References:

Ayinla, O. A. (2004). Integrated fish farming: A veritable tool for poverty alleviation/Hunger eradication in the Niger Delta Region. http://hdl.handle.net/1834/18746

Bajwa, A. A., Mahajan, G. & Chauhan, B. S. (2015). Nonconventional weed management strategies for modern agriculture. *Weed Science*, *63*(4), 723-747.

Berg, H. & Tam, N. T. (2012). Use of pesticides and attitude to pest management strategies among rice and rice-fish farmers in the Makeng Delta, Vietnam. *International Journal of Pest Management*, *58*(2), 153-164.

Debnath, C., Sahoo, L., Yadav, G. S., Datta, M. & Ngachan, S. V. (2013). Integrated fish farming. *Indian Farming*, *62*(10), 18-21.

Debnath, C., Sahoo, L., Yadav, G. S., Datta, M. & Ngachan, S. V. (2013). Integrated fish farming. *Indian Farming*, *62*(10), 18-21.

Ding, W., Xu, X., He, P., Ullah, S., Zhang, J., Cui, Z. & Zhou, W. (2018). Improving yield and nitrogen use efficiency through alternative fertilization options for rice in China: A meta-analysis. *Field Crops Research*, *227*, 11-18.

Ghosh, S. K., Mandal, B. K. & Borthakur, D. N. (1984). Effects of feeding rates on production of common carp and water quality in paddy-cum-fish culture. *Aquaculture*, *40*(2), 97-101.

Gomes, F. B. S. (2003). Developing a protocol for bottom soil and water quality management in ponds for semi-intensive culture of black tiger prawn *Penaeus Monodon*. Auburn University.

Goswami, M., Biradar, R. S. & Sathiadhas, R. (2004). Techno economic viability of rice fish culture in Assam. *Aquaculture Economics and Management*, *8*(5-6), 309-317.

Gupta, N., Singh, M. and Sharma, R. P. (2015). Evaluation of integrated horticulture-cum-fish farming in Malwa region of Madhya Pradesh, India. *Current World Environment*, *10*(**2**), 667.

Halwart, M. & Gupta, M. V. (2004). Culture of fish in rice fields. FAO; World Fish Center.

Haobijam, J. W., Balkho, M. R. & Mandal, T. K. (2018). Paddy-cum-fish cultivation and the challenges faced by the farmers of Manipur, India. *International Journal of Current Microbiology and Applied Sciences*, *7*(2), 998-1004.

Hossain, S. T., Sugimoto, H., Ahmed, G. J. U. & Islam, M. R. (2005). Effect of integrated rice-duck farming on farmers' rice yield, farm productivity, and rice-provisioning ability. *Asian Journal of Agriculture and Development*, *2*(1362-2016-107647), 79-86.

https://agritech.tnau.ac.in/animal_husbandry/animhus_ifs_pig.html

https://www.agrifarming.in/rice-fish-farming

https://www.google.com/search?client=firefoxd&sca_esv=4cf36d7972dd733c&q=fish+cum+poultry+farming&udm

Maucieri, C., Nicoletto, C., Zanin, G., Birolo, M., Trocino, A., Sambo, P. & Xiccato, G. (2019). Effect of stocking density of fish on water quality and growth performance of European Carp and leafy vegetables in a low-tech aquaponic system. *PloS one*, *14*(**5**), 0217561.

Mirhaj, M., Razzak, M. A. and Wahab, M. A. (2014). Comparison of nitrogen balances and efficiencies in rice cum prawn vs. rice cum fish cultures in Mymensingh, Northeastern Bangladesh. *Agricultural Systems*, *125*, 54-62.

Misra, V. K., Singh, C. P., Singh, N. D., Mishra, T. S., Mishra, N. K., Tripathi, A. N. and Singh, S. (2019). Integrated fish cum poultry farming for self employment and household nutritional security in Arunachal Pradesh. *Journal of Krishi Vigyan*, *7*(2), 163-167.

Mutua, B. M. (2018). Challenges facing indigenous chicken production and adoption levels of biosecurity measures in selected areas of Makueni County, Kenya (Doctoral dissertation).

Nair, C. M., Salin, K. R., Pillai, D., Sarath, S. & Manoj, C. K. (2010). Simultaneous Rice–Fish Culture System in Modified Pokkali Rice Fields–A Possible Alternative to Improve Sustainability. *B. Meenakumari, MR Boopendranath, Leela Edwin, TV Sankar, Nikita Gopal & George Ninan): Coastal Fishery Resources of India-Conservation and Sustainable utilization, Society of Fisheries Technologists (India), Central Institute of Fisheries Technology, Kochi, India.*

Nayak, P. K., Nayak, A. K., Panda, B. B., Lal, B., Gautam, P., Poonam, A. and Jambhulkar, N. N. (2018). Ecological mechanism and diversity in the rice-based integrated farming system. *Ecological Indicators*, *91*, 359-375.

Ngugi, C. C., Bowman, J. R. & Omolo, B. O. (2007). A New Guide to Fish Farming in Kenya.

Oben, B. O., Molua, E. L. & Oben, P. M. (2015). Profitability of small-scale integrated fish-rice-poultry farms in Cameroon. *Journal of Agricultural Science*, *7*(11), 232.

Pandey, A., Dhawan, A., Srivastava, A. & Singh, A. (2016). Integrated fish-cum-duck farming: productivity enhancement through technological interventions. *Indian Journal of Ecology*, *43*, 894-896.

Paul, I., Mandal, L. & Datta, S. (2018). Feasibility study of the small-scale poultry-cum-fish farming in homestead ponds for empowerment of rural women: A case study in the new alluvial zone of west Bengal, India. *International Journal of Fisheries and Aquatic Studies*, 6(3), 118-122.

Popp, J., Váradi, L., Békefi, E., Péteri, A., Gyalog, G., Lakner, Z.& Oláh, J. (2018). Evolution of integrated open aquaculture systems in Hungary: results from a case study. *Sustainability*, *10*(1), 177.

Rai, S. C. (2005). Apatani paddy-cum-fish cultivation: An indigenous hill farming system of North East India.

Saha, R. K. & Nath, D. (2013). Indigenous Technical Knowledge (ITK) of fish farmers at Dhalai district of Tripura, NE India.

Sastry, K. S., A. Zitter, T., Sastry, K. S. & Zitter, T. A. (2014). Management of virus and viroid diseases of crops in the tropics. *Plant Virus and Viroid Diseases in the Tropics: Epidemiology and Management*, 2,149-480.

Yani, P. & Sharma, A. (2018). Priortization strategies for the resources of traditional paddy-cum-fish culture in lower subansiri district of Arunachal Pradesh. *Journal homepage: http://www. ijcmas. com*, 7(05), 2018.

CHAPTER 7

COMPOSITE FISH FARMING

Introduction

The composite fish culture system operates on the core concept of stocking diverse, fast-growing fish species that harmonize well with each other and exhibit complementary feeding behaviors. Its objective is to maximize fish yield by efficiently utilizing the natural food resources in various ecological niches within the pond. To implement this technology, several crucial steps must be followed. Initially, the pond undergoes the removal of aquatic weeds and predatory fishes to establish a conducive environment for the selected fish species. Liming is then conducted, and appropriate fertilizers tailored to the soil and water quality of the pond are applied. Subsequently, the pond is stocked with 100 mm-sized fingerlings of Indian major carp species like catla, rohu, and mrigal alongside exotic carp varieties such as silver carp, grass carp, and common carp. The selection and density of these fish species are meticulously planned to ensure optimal growth and productivity. Supplementary feeding forms an integral part of the composite fish culture system, providing additional nutrition to enhance the fish's growth rate alongside their natural diet. Harvesting is carried out at the right time to achieve the desired yield. The composite fish culture system offers three primary combinations: the sole culture of Indian major carp, the sole culture of exotic carp, and the combined culture of both Indian and exotic carp. Each combination presents its unique advantages and considerations. Typically, fish production ranging from 3,000 to 6,000 kg per hectare annually can be attained through this system. However, advancements in intensive pond management have led to further enhancements in fish yield. Recently, integrated fish and animal husbandry systems have gained prominence. These include fish-cum-duck culture, fish-cum-poultry culture, fish-cum-pig culture, and the utilization of cattle farm yard wastes and biogas plant slurry recycling for fish production. These integrated systems offer additional benefits and contribute to productivity and sustainability (Talukdar and Sontaki, 2005).

The benefits of integrating culture systems, bird/animal population assessment, manure quantity determination, and the fish yield potential of recycling setups are elucidated. Fish cultivation in paddy fields emerges as a crucial symbiosis of aquaculture and agriculture. The fundamental requisites for pisciculture in paddy fields, the desirable attributes of rice field culture, challenges arising from contemporary farming practices, and advancements in fish-paddy cultivation methods are thoroughly examined. The burgeoning trend of freshwater prawn farming, mainly focusing on *Macrobrachium rosenbergii* (giant freshwater prawn) and *M. malcolmsonii* (Indian riverine prawn) as the preferred species in India, is explored. Comprehensive discussions cover breeding techniques, hatchery management, seed production, various culture methodologies, and the production potential of freshwater prawns. Significant air-breathing fish species in India, such as murrels, climbing perch, singhi, and maguro, are highlighted, accompanied by insights into seed production methodologies and optimal culture systems (Biswas *et al.*, 2018).

Composite Fish Culture:

- Aquaculture aims to optimize fish yield from aquatic environments such as ponds.
- Techniques for raising fish involve carefully managing environmental factors like soil, water quality, and the fish population's health.
- Key aspects that aid in successful fish farming include minimizing water usage by the fish and ensuring high reproductive rates.
- Fish are recognized as a valuable food source, offering a rich supply of protein, essential vitamins, and nutrients crucial for human nourishment and development. The growing population and limited cultivable lands suggest that animal protein availability may decrease.
- Therefore, it becomes necessary to obtain more animal protein from aquatic sources.
- We must explore ways to increase fish production to meet the increasing demand for animal protein.
- Controlled pond conditions offer the most significant potential for producing more fish.
- The fish pond functions as a complex ecosystem with various components and interactions. Floating organisms such as phytoplankton and zooplankton can be found at the surface. In the column region of the pond, there's a blend of living and deceased organic material that gradually descends from the surface. The pond bed is notably enriched with detritus, constituting dead organic matter. Along the pond's margins, there's a rich diversity of aquatic vegetation, as outlined by Swetha *et al.* (2018).
- To enhance the profitability of fish culture, the different trophic levels within the pond are effectively utilized. A recent approach to fish culture known as

composite fish culture, polyculture, or mixed farming has been developed. The main goal of this concentrated fish farming approach is to identify and nurture fish species that can coexist peacefully with varied eating habits. This method enables the utilization of all food resources within different areas or ecological niches of the fishpond, leading to the highest possible fish yield.

- In earlier times, the amount of fish harvested from ponds was relatively low, typically around 500 kg per hectare per year, which was deemed unsatisfactory. Nevertheless, through composite fish farming techniques, it has become feasible to attain yields surpassing 10,000 kilograms per hectare per year in various agro-climatic zones across our nation (Hussain *et al.*, 2013).

Superiority over the monoculture

Monoculture is cultivating only one type of fish in a pond. When a single species is introduced into the pond, they tend to congregate in one area due to their similar eating habits. Opting for monoculture involves introducing many fish from the same species, leading to intense competition for food and space. This heightened competition often results in fights among the fish, leading to a significant mortality rate. Additionally, the fish do not receive enough nutrition due to the limited food supply, which hinders their growth potential. Consequently, the overall yield is negatively impacted. Furthermore, in monoculture systems, other ecological niches remain vacant, causing the available food in those niches to go to waste (Basavaraju and Reddy, 2013).

Composite fish culture is unquestionably superior to monoculture due to its numerous advantages. Unlike monoculture, composite fish culture eliminates the issues mentioned above. By incorporating six different varieties of fish, this system maximizes the utilization of food from all niches within the pond. Consequently, the fishes receive ample food, ensuring their healthy growth without competition. Moreover, the yield obtained from composite fish culture is significantly higher. The mortality rate in this system is negligible, further emphasizing its superiority over monoculture.

In contrast, monoculture struggles to achieve a yield of approximately 500 kg per hectare per year. However, as employed in composite fish culture, the polyculture system yields approximately 20 times more than monoculture when managed with scientific precision. These stark differences highlight the significant benefits and advantages of composite fish culture, making it the preferred approach for fish farming (Das, 2017).

Principles of composite fish culture

- The composite fish culture technique, rooted in scientific principles, strives to optimize pond productivity by fostering symbiotic relationships among various food fish species. This method entails co-cultivating fast-growing, non-predatory, and non-competitive fish species, considering their complementary feeding behaviours and capacity to utilize natural and supplemental food sources. The fish benefit from each other's presence through this integrated approach. For instance, the waste produced by grass carp serves as nourishment for fish food organisms, becoming a food source for other fish species. This symbiotic interaction minimizes competition for space and food resources among the fish.

- Furthermore, bottom-feeding species like common carp and marginal play a crucial role by consuming a portion of the waste generated by grass carp. Their presence is essential for preventing water pollution caused by excessive accumulation of grass carp waste. Therefore, maintaining an appropriate population of bottom feeders is vital for preserving ecological equilibrium within the pond. By carefully stocking an optimal number of each fish species, the diverse environmental niches within the pond are effectively utilized, as highlighted by (Kund *et al.*, 2010).
- To boost the pond's ability to support more fish, you can stimulate the production of natural fish food by using fertilization methods. Additionally, supplementary feed ensures enough food to help many fish stocked in the pond. This approach optimizes the utilization of available resources, leading to increased productivity and efficient fish farming practices (Biswas *et al.*, 2018).

Fishes used in composite culture:

Around the globe, the predominant fish species suitable for cultivation, especially for polyculture, come from the carp family. There are three primary methods of carp farming practiced worldwide:

1. **Chinese method:** This approach involves cultivating various types of Chinese carp together, including silver carp (*Hypophthalmichthys molitrix*), grass carp (*Ctenopharyngodon idella*), and common carp (*Cyprinus carpio*). These carp species are also known as exotic fishes in India.
2. **Indian method**: In this system, Indian carp species like rohu (*Labeo rohita*), catla (*Labeo catla*), and mrigal (*Cirrhinus mrigala*) are cultured together, sometimes alongside Chinese carp.
3. **European method:** The main focus of the European method is cultivating the common carp (*Cyprinus carpio*).

In addition to these primary species, other Chinese carp like big-head carp (*Aristichthys nobilis*), mud carp (*Cirrhinus molitorella*), and black carp (*Mylopharyngodon piceus*) are utilized in composite fish culture.

Introducing predatory catfish and murrels into the composite fish culture system is feasible. Still, teaching them only after the carp species have reached a significant size is crucial. Trash fish and young common carp in the culture pond can serve as nutritious food sources for the catfish and murrels.

In brackish water culture systems, fringe-lipped carp and milkfish are commonly cultivated alongside other species in composite fish culture. Furthermore, air-breathing fishes such as murrels, catfishes, and koi are often cultured together in freshwater culture systems.

The concept of composite and integrated fish farming differs between regions. In India and China, polyculture, which involves raising multiple species together, is more popular. On the other hand, European countries tend to practice monoculture, where a single species is cultivated. The dominance of common carp as a cultivated species worldwide can be attributed to the relative ease of seed production compared to other cultivable carp.

Indian significant carp predominantly inhabit rivers and tend not to breed in enclosed bodies of water. Consequently, their offspring are typically harvested during the monsoon season when river flooding occurs. Identifying specific species during natural collection is complex, often resulting in a mixture of undesired species being stocked in ponds alongside the desired ones. Over time, this practice gave rise to polyculture, a method of cultivating multiple species simultaneously. Recently, the scientific underpinnings of polyculture have gained recognition (Kant *et al.*, 2013).

In the late 1950s, India introduced exotic carp varieties, including the common carp, silver carp, and grass carp. Since then, these species have been effectively cultivated together, often alongside native Indian major carp. Including grass carp in these aquaculture systems is particularly vital, as it is crucial in biologically managing aquatic weeds. Grass carp have a voracious appetite for marine vegetation. Composite fish culture, which involves raising multiple fish species together, is a significant advancement in freshwater aquaculture in the country. This technology evolved during this period, particularly in stocking ponds with different species.

Much of the initial energy diminishes in a food chain as it progresses through each trophic level. Thus, practical fish farming endeavours to minimize the food chain's length. Herbivorous fish are favoured alongside those that consume zooplankton. Typically, predatory fish are best kept out of the system.

A mixture of plankton and macrophyte feeders is expected to be stored in fish culture systems. These organisms rely on the nutrients already in the ponds or those applied from external sources. However, maintaining a proper balance is crucial, as without it, one group may outcompete the other.

To maintain a harmonious ecosystem, stocking involves a variety of fish with diverse feeding behaviours. Zooplankton, which consume ungrazed phytoplankton, are crucial to this balance. Therefore, it's important to include fish species that feed on zooplankton. In India, the suggested mix for a polyculture system includes rohu, catla, mrigal, common carp, silver carp, and grass carp. These species possess unique feeding patterns, don't vie for resources, and aren't predatory (Uddin, 2007).

Rohu is a fish that mainly eats plankton, dwelling in its environment focusing on column-fed sources. On the other hand, Catla specializes in feeding on surface-dwelling zooplankton. Mrigal is known as a bottom feeder, preferring to consume plankton and benthic organisms found at the pond's bottom. The common carp also feeds at the bottom, preferring detritus. Silver carp, unlike the others, feeds exclusively on phytoplankton at the water's surface. Grass carp, however, primarily feed on aquatic vegetation, using a substantial portion of the pond's available food organisms.

In China and Southeast Asia, the most common combination consists of silver carp, which feeds on phytoplankton; big head, which feeds on zooplankton; and grass carp, which consumes aquatic vegetation. This combination ensures the utilization of various food sources and promotes a balanced ecosystem in the fish culture system (Rahman *et al.*, 2009).

Stocking densities and stoking ratio

Generally, when more fish are put into an area, the amount of fish caught tends to go up until a point, but then it starts going down. There's an ideal number of fish to put in an area to get the most fish caught and the biggest.

When fish are stocked at a higher density, especially in crowded conditions, they may engage in aggressive interactions and compete intensely for food. This can lead to stress among the fish, causing them to eat less and grow slower. Enhancing the quantity of fish stocked beyond the ideal level not only heightens the need for oxygen, presenting evident risks but also does not contribute to an overall increase in fish yield. When establishing the stocking density and proportion of fish, it is vital to account for the available water volume and oxygen generation. For the six types of Indian and Chinese major carp specified earlier, a suggested stocking rate is 5000 fingerlings of 75-100 mm size per hectare. As mentioned earlier, the allocation of stocking percentages for each of the fish should be determined based on the specific requirements and recommendations of those species (Das & Mishra 2016).

- Catla and Silver carp -30-20%
- Rohu – 15-20 %
- Mrigal and common carp -45%
- Grass carp – 5-10 %

In a combination of five species, with grass carp excluded, the recommended stocking ratios are as follows: Catla 6 (30%), Rohu 3 (15%), Mrigal 5 (25%), Common Carp 4 (20%), and silver Carp 2 (10%).

For a combination of four species, excluding both silver carp and grass carp, the optimal stocking ratios would be: Catla 6 (30%), Rohu 3 (15%), Mrigal 6 (30%), and Common Carp 5 (25%).

In a three-species combination excluding exotic carp, the suggested ratios are Catla 4 (40%), Rohu 3 (30%), and Mrigal 3 (30%).

In composite fish culture involving an eight-species combination, including milkfish and fringe-lipped carps alongside significant Indian and Chinese carps, it is feasible. However, the growth of the additional species is not satisfactory. The stocking ratio for this combination typically consists of Catla 2, Rohu 2, Mrigal 4, Common Carp 3, Silver Carp 5, Grass Carp 2, Fringe-lipped Carp 1, and Milkfish Note: It's important to note that milkfish is a brackish water fish (Ansal *et al.*, 2016).

Feeding

With the pond's carrying capacity boosted, whether through aerating the water or fostering fish growth, additional fish production can be attained by introducing supplementary feed. To achieve optimal production levels, fish are given feed rich in protein. Typically, the conversion coefficient is 1:2, meaning that 2 kilograms of feed is provided for every 1 kilogram of fish yield. The fish's growth can increase tenfold by incorporating supplementary feeds such as rice bran and oil cake.

Grass carp primarily feed on tender aquatic plants such as *Najas, Hydrilla, Ceratophyllum*, and *Chara*, along with forage grasses or chopped green cattle fodders like Napier grass, Barseem, maize leaves, etc., as well as kitchen vegetable scraps. The cattle fodder is grown on the pond's terraced embankment and provided to the grass carp. Feeding occurs twice daily, starting at 100 kilograms per hectare in the initial month and increasing by 100 kilograms monthly at fortnightly or monthly intervals until the end of the harvesting period. To facilitate feeding, a floating frame made of bamboo poles is typically used to hold the food for the grass carp.

Harvesting and yield

The general recommendation for harvesting fish is to wait for approximately one year of rearing. However, shorter rearing periods may also be considered depending on the pond's conditions and the local market's size preference. Over a year, a single fish usually attains a size ranging from 0.8 to 1 kg. In contrast, grass carp exhibit accelerated growth, achieving weights of up to 3 kilograms annually. They make up roughly 30% of the overall fish yield in a pond. A recent milestone in Pune highlighted a breakthrough in fish production using composite farming methods. In a pond spanning 0.31 hectares and stocked with 8,000 fingerlings per hectare, an impressive output of 10,194 kg/ha/year was obtained. It is easily feasible to achieve an average production of 5,000 kg/ha/year using this culture system, underscoring the significant potential for fish production through composite fish culture (Majumder and Saikia 2020).

Trial netting is conducted monthly to assess the fish's development and growth. This practice also aids in the prompt identification of any parasitic infections that may occur. Moreover, meshing proves advantageous in purging the pond floor, expelling toxic fumes from the pond bed and removing nutrients from the underlying terrain.

In a trial concerning the mixed cultivation of brackish water fish species like *Chanos chanos, Mugil cephalus, Etroplus suratensis*, and *Liza parsia*, a yield rate of 2189 kg per hectare per annum was attained. The amalgamation of Chanos and Mugil displayed the highest output, with Chanos exhibiting superior proliferation followed by Mugil (Hielmstedt, 2022).

Hazards in composite fish culture

Composite fish farming faces hazards that can lead to significant losses if not addressed promptly. These hazards can be categorized as biological, management-related, or harvesting issues. It is crucial to anticipate these risks and implement remedial measures to mitigate them effectively. Poor management practices are often responsible for most problems in composite fish farming (Hiiam *et al.*, 2015).

Biological problems

From the pond to prevent them from preying on the fish. Also, snakes are potentially dangerous as they can attack and consume the fish. Therefore, it is crucial to implement appropriate measures to mitigate these biological hazards.

To effectively manage the proliferation of aquatic vegetation, it is strongly advised to introduce herbivorous fish such as grass carp and *Puntius* varieties. These fish are recognized for their ability to consume and diminish weed populations within ponds. Additionally, several predatory fish species such as *Mystus, Ompok, Wallago, Notopterus, Oreochromis,* and *Gobius,* along with weedy species like *Salmostoma, Esomus, Barbus, Ambasis, Rasbora,* and *Amblypharyngodon,* are frequently present in pond ecosystems and may pose competition to carp fingerlings. It is essential to eradicate these predatory and weedy fish species during the pond's preparation to ensure the optimal growth and survival of the stocked fish.

Aquatic insects like beetles (*Cybister, Stemolopus*), bugs (*Belostoma, Anisops*), and dragonfly nymphs present a possible danger to the fish population alongside fish species. Therefore, it is necessary to eliminate these insects from the pond to prevent them from causing harm. The risks associated with these biological hazards can be significantly reduced by taking appropriate eradication measures, promoting a healthier and more productive environment for cultured fish.

Snakes and other organisms can inflict significant harm to fish crops as they prey on fingerlings. Another detrimental factor is the presence of numerous molluscs, which consistently negatively impact fish. However, this issue can be mitigated by introducing *Pangasius* into the pond. These fish can consume molluscs, thus decreasing their population and minimizing the extent of infestation (Jayasankar *et al.*, 2014).

Due to the precocious development and inherent reproductive characteristics of the common carp, their proliferation is accelerated, necessitating the implementation of precautionary steps to regulate the stocking density within the culture pond. Consequently, harvesting the common carp before reaching full maturity is imperative. An alternative approach involves the deliberate presence of aquatic weeds in designated areas of the pond, which serve as breeding grounds for the carp, as their eggs possess adhesive properties. These weeds and the attached eggs can subsequently be extracted from the pond, enabling the possibility of separate incubation to acquire viable hatchlings, if desired. Common carp, because of its burrowing nature, can spoil the dyke by making holes in it. Crabs also damage the dyke. Tilapia is a continuous breeder; hence, it must be avoided in the ponds.

Algal blooms like *Microcystis*, *Euglena*, etc., occur mainly in summer and pose significant issues with dissolved oxygen levels. Oxygen becomes supersaturated during the day and depleted at night. The most effective way to control these blooms is through chemical methods. In an emergency, pumping freshwater into the pond is a safe approach. To prevent sunlight penetration, a portion of the pond can be covered with shade-providing plants such as *Eichornia* and *Pistia*. However, if these plants spread within the pond again, eliminating them becomes a major challenge.

The main hazard is low oxygen levels in the water. Fish swim near the surface, gasping for air. It affects their growth and can lead to mass deaths. To save the fish, the farmer adds freshwater to the pond and beats the water with bamboo poles to increase oxygen. $KMnO_4$ (1ppm) can be added to boost dissolved oxygen and disinfectant. Quick-lime or slake-lime (200 Kg/ha) helps counteract the adverse effects of decaying organic matter. Dragging a net repeatedly releases harmful gases. Using cut banana stems can benefit the fish in this situation.

In composite fish farming, the proliferation of plant material is effectively controlled by the presence of silver carp and grass carp. These carp species sustain themselves by consuming phytoplankton and aquatic weeds, respectively. Additionally, including mrigal and common carp significantly mitigates the detrimental consequences arising from oxygen depletion caused by the decomposition of organic matter, as these fish feed on it. It's important to highlight that tall trees and bamboo vegetation cover many ponds within the village. This creates a notable challenge for photosynthesis in the ponds as it limits sunlight reaching the water. Ideally, it's best to minimize the presence of trees and bamboo around the pond edges. Banana plants can be strategically placed on the embankment, avoiding the eastern side to ensure uninterrupted morning sunlight. Additionally, it's advised to maintain banana plantations in a non-bushy manner (Tripathi and Ranadhir, 1982).

Management problem

The pond must always have at least one meter of water in it. Severe drought harms the water level in the ponds fed by rainwater. Alternative water supplies, such as tube wells, may be available. Some assistance in the drought fight. By bursting their dykes or flooding them to the point of overflow, heavy rain and floods seriously harm the ponds. In both situations, fish get out of the pond. It may be necessary to take short-term measures like pond screening or dyke protection.

In some cases, it is preferable to harvest the fish even before such a circumstance arises. Another issue in fish culture is poaching. In addition to hiring guards, bushy plant materials can be added to the ponds to deter simple netting. Trained watchdogs may be more cost-efficient and effective at stopping poaching (Farhana, 2011).

Harvesting problem

To ensure optimal utilization of resources and maintain the health of fish stocks, it is crucial to harvest them before their growth rate declines in response to invested inputs like feed and fertilizers. Once the fish reach a particular stage, enhancing the nutritional value of the water feeding them becomes impossible. The harvesting process becomes more complex due to differential growth rates among the fish. To address this challenge, it is recommended that partial harvesting can be considered a viable option if the timing of harvesting proves to be highly challenging to synchronize within a fish community, despite careful manipulation of stocking ratios and density.

The pricing of fish weighing less than a kilogram is generally lower compared to those weighing over a kilogram or more. This factor also has an impact on the fishing schedule, and considering this aspect is crucial for maximizing profits before harvesting (Abakumov and Izrailsky 2022).

It is also important to seriously consider the interrelationship among the cultured species. Bottom feeders partially rely on the fecal matter of grass carp for sustenance. Therefore, an unplanned removal of grass carp would subsequently affect the growth of the bottom feeders. On the other hand, if only bottom feeders are completely harvested, the excessive fecal matter from the grass carp could potentially pollute the water. Although there are hazards associated with composite fish culture, they can be effectively managed and mitigated through proper precautions and vigilant measures.

Economics

The financial dynamics of composite fish culture can fluctuate significantly across different regions, mainly influenced by land-value, soil quality, labour expenditures, construction material costs, and transportation charges. The economic framework of fish farming and its related cost structures are not easily standardized due to these diverse variables.

However, the profitability of fish production is generally considered high overall. Despite the variations in production economics, composite fish culture systems have proven financially rewarding for farmers. This profitability results from various factors, including high demand for fish in the market, efficient production techniques, and the ability to optimize resources and achieve high yields.

It is important to note that the economics of fish production in composite fish culture are site-specific and depend on local conditions. Land prices can differ significantly between regions, impacting the overall investment required. Soil conditions also play a vital role in affecting water quality, nutrient availability, and fish growth. Labour costs, including wages and availability of skilled workers, can vary depending on the location.

Additionally, the cost of farm construction materials, such as ponds, tanks, or infrastructure, can differ based on the region and availability of resources. Transportation costs, including the proximity to markets and infrastructure, also affect the economics of fish production.

Considering all these factors, it is evident that the economics of fish production in composite fish culture is complex and dynamic, making it challenging to generalize the nature of production and its cost functions. Nonetheless, the sector remains highly profitable, offering opportunities for farmers who can adapt their production systems to local conditions and effectively manage their resources (Brown, 2012).

References

Abakumov, A. & Izrailsky, Y. (2022). Optimal harvest problem for fish population structural stabilization. *Mathematics, 10*(6), 986.

Ansal, M. D., Dhawan, A., Singh, G.& Kaur, K. (2016). Species selection for enhancing the productivity of freshwater carps in InlandSaline Water of Punjab-A Field Study. *Indian Journal of Ecology, 43*(1), 45-49.

Basavaraju, Y. & Reddy, A. N. (2013). Growth performance of Amur strain of common carp in southern Karnataka. *Mysore Journal of Agricultural Sciences, 47*(1), 119-123.

Biswas, B., Das, S. K. & Mandal, A. (2018). Socio-economic dimensions and their impacts upon the productivity of composite fish farming in north 24 parganas district, West Bengal. *Nature, 25*(35), 41-67.

Biswas, B., Das, S. K., Mondal, I. & Mandal, A. (2018). Composite fish farming in West Bengal, India: redesigning management practices during the course of last five decades. *International Journal of Aquaculture, 8*(12).

Brown, E. (2012). *World fish farming: cultivation and economics*. Springer Science and Business Media.

Das, P. C. & Mishra, B. (2016). Multi-species farming of major and minor carps for enhancing fish production in freshwater aquaculture. *Indian Journal of Fisheries, 63*(2), 55-61.

Das, S. K. (2017). Performance of an improved breed of common carp-Amur (Hungarian strain) in the North-eastern hill state of Meghalaya, India. *Indian Journal of Fisheries. Indian Journal of Fisheries., 64,33-38, 2017 DOI: 10.2107/ijf. 2017.64.*

Farhana, S. (2011). Suitability of pond sand filters as safe drinking water solution in storm surge prone areas of Bangladesh: a case study of Post-Aila situation in Shyamnagar, Satkhira district, Khulna (Doctoral dissertation, BRAC University).

Hijam, B., Pandey, D. K. & De, H. K. (2015). Training Need of Fish Farmers in Bishnupur District of Manipur: An Analysis. *Indian Journal of Extension Education, 51*(34), 152-155.

Hjelmstedt, P. (2022). Safeguarding the welfare of fish in aquaculture: physiological assessments of stress and welfare during handling, transport and slaughter.

Hussain, S. M., Sen, D. E. B. A. S. H. I. S. H., Pathak, M. A. H. E. S. H. & Singh, M. P. (2013). Comparative study of composite fish culture (CFC) and local practices of fish culture in East Siang District, Arunachal Pradesh. *Indian Journal of Hill Farming, 26*(2), 32-34.

Jayasankar, P., Sahu, B. B., De, H. K., Rajesh, N., Panda, N., Mohanty, U. L.& Sahoo, P. R. (2014). Problems in aquaculture development in Odisha a participatory assessment.

Kant, R., Haq, S., Srivastava, H. C. and Sharma, V. P. (2013). Review of the bioenvironmental methods for malaria control with special reference to the use of larvivorous fishes and composite fish culture in central Gujarat, India. *Journal of Vector Borne Diseases, 50*(1), 1.

Kund, G. C., Mishra, R. and Sethi, P. K. (2010). An economic analysis of composite fish culture ponds in Sundergarh district, Orissa. *Asian Journal of Animal Science, 5*(2), 139-141.

Majumder, S. & Saikia, S. K. (2020). Ecological intensification for feeding rohu *Labeo rohita* (Hamilton, 1822): A review and proposed steps towards an efficient resource fishery. *Aquaculture Research, 51*(8), 3072-3078.

Rahman, M. M., Hossain, M. Y., Jo, Q., Kim, S. K., Ohtomi, J. and Meyer, C. (2009). Ontogenetic shift in dietary preference and low dietary overlap in rohu (*Labeo rohita*) and common carp (*Cyprinus carpio*) in semi-intensive polyculture ponds. *Ichthyological research, 56*, 28-36.

Swetha, M., Maloth, M., Naaiik, R. V. T., Rajkumar, B., Kumar, P. V., Bhavyamanjari, M. & Suresh, M. (2020). Impact of training programme on knowledge level of fish farmers on composite fish culture. *Current Journal of Applied Science and Technology, 39*(41), 1-10.

Talukdar, P. K. & Sontaki, B. S. (2005). Correlates of adoption of composite fish culture practices by fish farmers of Assam, India.

Tripathi, S. D & Ranadhir, M. (1982). Economic analysis of composite fish culture in India. In: *Aquaculture Economics Research in Asia: proceedings of a workshop held in Singapore, 2-5 June 1981*. IDRC, Ottawa, ON, CA.

Uddin, M. S. (2007). Mixed culture of tilapia *(Oreochromis niloticus)* and freshwater prawn *(Macrobrachium rosenbergii)* in periphyton-based ponds. *Wageningen University and Research.*

CHAPTER 8

IMPACT OF INORGANIC MINERALS AND IONS ON AQUACULTURE PRODUCTIVITY

Introduction

Minerals are naturally occurring, inorganic substances with a definite chemical composition and a specific atomic arrangement. They are solid and have a crystalline structure, meaning their atoms are arranged in a repeating pattern. As John (2011) outlined, minerals are typically produced through geological processes over extended periods. These inorganic nutrients, essential for various bodily functions, are usually needed in small quantities ranging from less than 1 to 2500 mg per day, depending on the specific mineral, as indicated (Soetan *et al.*,2010). When an element is labelled as "dietary essential," it signifies that its absence or insufficient intake in the diet can lead to deficiency syndrome, and restoring it to normal levels through supplementation promotes overall health. These essential elements cannot be substituted wholly or partially by any other elements. Aquatic organisms, including crustaceans, fish, and aquatic plants involved in aquaculture, as noted by FAO (2020) and Tacon (2022), require certain inorganic elements in minimal or trace amounts to support their average growth and metabolic functions. Optimal levels of trace elements are necessary to enhance the productivity and well-being of aquatic animals, as emphasized (Dawood *et al.*, 2021). It's important to note that trace minerals often exist in conjunction with others, and excessive intake of one element may disrupt the balance of others, potentially leading to diseases or adverse effects on metabolism. An optimized diet for each reared species is the ultimate goal of sustaining productivity and profitability, so formulation and feed management play an essential role in the aquaculture industry. The use of growth stimulants in manufactured feeds has received attention in the past years for enhancing fish growth performance and production (Hassan *et al.*, 2021). Factors such as dietary intake, absorption capacity, potential toxicities, and interactions with medications may influence the maintenance of trace mineral balance in animals (Glover *et al.*, 2004).

In contrast to livestock, the understanding of the dietary necessity of minerals in fish is limited, primarily focusing on iron, copper, manganese, zinc, iodine, selenium, Cobalt, and chromium. These minerals serve various functions in fish, including as components of bodily fluids, cofactors in enzymatic processes, and structural elements of non-enzymatic molecules. Despite being required in small amounts, trace minerals are recognized as essential components in fish diets (Chanda *et al.*, 2015). Fish have been found to require approximately ten minerals for optimal health, but the need for fish depends on water resources and their salinity level. Based on salinity, three types of resources are present (Loitzenbauer and Mendes, 2012). Calcium is essential for skeletal development and muscle function. Copper is necessary for enzyme activities and iron metabolism. Iodine is crucial for thyroid hormone synthesis. Iron is necessary for oxygen transport and various metabolic processes. Magnesium contributes to enzyme function, while manganese is a cofactor for several enzymes. Phosphorus is a critical component of nucleic acids and energy metabolism. Potassium is involved in osmoregulation and muscle contractions. Selenium serves as an antioxidant and is involved in thyroid function. Zinc plays a role in enzyme function, growth, and immunity.

Definition

Macrominerals refer to essential minerals adults need in quantities exceeding 100 mg per day. This group comprises sodium, potassium, chloride, calcium, magnesium, and phosphorus. Trace elements, on the other hand, are minerals required in adult diets ranging from 1 to 100 mg per day. Examples of trace minerals are iron, copper, and zinc. Ultra-trace minerals are those needed in quantities less than 1 mg per day, such as chromium, manganese, fluoride, iodide, Cobalt, selenium, silicon, arsenic, boron, vanadium, nickel, cadmium, lithium, lead, and molybdenum.

Classification:

Minerals are classified into two major groups: one required in relatively large amounts (macrominerals) and the other in tiny amounts (microminerals). The macro minerals are calcium, phosphorus, potassium, sodium, chloride, and magnesium, while microminerals can include chromium, Cobalt, copper, fluoride, iodine, iron, molybdenum, manganese, selenium, sulfur and zinc (NRC, 2011; Webster and Lim, 2015)

Table. 1 Classification of macrominerals and microminerals

Sr. No.	Macrominerals	Microminerals
1.	Calcium (Ca)	Cobalt (Co)
2.	Phosphorous (P)	Iron (Fe)
3.	Chlorine (Cl)	Copper (Cu)
4.	Magnesium (Mg)	Manganese (Mn)
5.	Potassium (K)	Selenium (Se)
6.	Sodium (Na)	Zinc (Zn)
7.		Chromium (Cr)
8.		Molybdenum (Mo)
9.		Sulphur (S)
10.		Flluorine (F)
11.		Iodine (I)

(Source: Authors)

Impact of ions on water quality

The concentration of substances in a recirculating aquaculture system (RAS) can impact common carp's embryonic and larval growth. In a study by Martins *et al.* (2019), it was observed that water from a RAS operating at 30 L/kg feed/day, known as high-accumulation water, led to increased mortalities during both embryonic and larval stages compared to water with lower substance accumulation. Additionally, hatching percentages were lower in high-accumulation water compared to low-accumulation water. Surprisingly, despite the higher mortality rate, larvae hatched in low-accumulation water exhibited a higher incidence of deformities than those in high-accumulation water. Unionized ammonia is recognized as highly toxic to fish. However, in the study, the level of un-ionized ammonia did not significantly differ between treatments. The highest observed NH_3 value (0.007 mg/L) was ten times lower than the lowest concentration known to affect early-life developmental stages in fish, according to Fairchild *et al.* (2005).

Nitrate levels ranging from 1.1 to 4.5 mg NO_3^- N/L (Nitrate nitrogen) are considered mildly toxic to early-life stages of various fish species such as rainbow trout, Chinook salmon, Steelhead trout, and cutthroat trout. McGurk *et al.* (2006) found the acute median lethal concentration (LC50) for swim-up fry to be 1.1 mg NO_3^- N/L for lake trout and 1.9 mg NO_3^- N/L for lake whitefish. Despite this, the exact mechanism through which high nitrate levels influence embryonic and larval development remains unclear. Some studies have suggested a link between nitrate toxicity and osmoregulation issues. Still, nitrate uptake into the plasma appears to be passive, with plasma concentrations remaining below ambient nitrate levels after eight days of exposure. Furthermore, it has been noted that the limited uptake of nitrate does not seem to affect electrolyte balance or haematology in juvenile rainbow trout.

Note: Lethal concentration 50 (LC_{50}) is the amount of a substance suspended in the air required to kills 50% of a test animals during a predetermined observation period

Effect:

The limited availability of specific ions for versatile saltwater crustaceans like *L. vannamei* leads to increased energy expenditure to balance salt concentrations within their cells due to osmotic pressure (Boyd, 2015 and Boyd *et al.*2016). Additionally, various ions, such as calcium (Ca^{2+}) and magnesium (Mg^{2+}), play crucial roles in the metabolic, chemical, and physiological processes of L. vannamei, including facilitating the ecdysis process. Sodium (Na^{+}) and potassium (K^{+}) are essential for growth, osmoregulation, and immune functions (Katya *et al.*, 2016). In environments with low concentrations of these ions, *L. vannamei* cultures may experience heightened mortality rates and growth constraints (Saoud *et al.*, 2003; Roy and Davis, 2007).

Zinc

Zinc (Zn), a crucial micronutrient (Watanabe *et al.*, 1997), ranks second in abundance among trace elements, following Fe. It is indispensable for cellular functions across most organisms, playing pivotal roles in various physiological processes in fish. These include enzyme activities, cellular structure maintenance, hormone synthesis, and biological metabolism (Philip *et al.*, 2023). As the most prevalent intracellular trace element, Zn's widespread presence within cells has recognized three distinct functions in vertebrate biology. Firstly, Zn serves a catalytic function vital for the biological activity of over 300 enzymes across six enzyme classes, such as RNA nucleotide transferases (RNA polymerase I, II, and III), alkaline phosphatase, and carbonic anhydrases. Secondly, Zn plays a structural role in metalloenzymes, stabilizing their tertiary structures. For instance, Cu acts as the catalytic site in Cu-Zn superoxide dismutase, while Zn contributes to structural integrity.

Zn finger motifs in proteins featuring histidine instead of cysteine are crucial for structural stability. Dietary imbalances of Zn, either deficiency or excess, can significantly impact fish growth and health. Hence, it is imperative to maintain optimal Zn levels in aquafeeds for cultured fish. For salmonids like salmon and trout, recommended Zn inclusion levels in the feed may range from 50 to 100 mg/kg, while for tilapia, it could range from 30 to 60 mg/kg. Like other trace elements, Zn primarily enters the fish's system through water, mainly via the gills and gastrointestinal tract. However, the gastrointestinal tract serves as the primary site for Zn absorption. Mechanisms to regulate excessive Zn uptake and accumulation in the gills are also in place, ensuring homeostasis.

Currently, numerous studies have been conducted to ascertain the dietary zinc requirements and investigate how different levels of dietary zinc influence growth, zinc content in tissues, and antioxidant capabilities in fish (Wei *et al.*, 2018; He *et al.*,

2023). Muscle growth is influenced by various factors, such as myogenic regulatory factors (MRFs), Pax 7, MEF, and myostatin. MRFs, including MyoD, MyoG, Mrf4, and Myf5, are pivotal transcription factors associated with muscle development, while MEF2 is a critical myocyte enhancer-binding factor. Research suggests an optimal dietary zinc level promotes growth performance and augments antioxidant and inflammatory responses in yellow catfish muscle. Additionally, the addition of dietary zinc influences the expression of genes related to muscle growth and development. Nonetheless, further investigation is necessary to elucidate how zinc affects these aspects of yellow catfish muscle (Xu *et al.*, 2023).

However, there is currently a lack of information regarding the impact of zinc on antioxidant and inflammatory signalling in fish muscle. Zn deficiency reduces protein and carbohydrate digestibility, thereby negatively affecting fish growth. By contrast, excess Zn in the diet may harm the fish and cause adverse effects on the nutritional status of other essential elements (Kucukbay *et al.*, 2006). According to the EPA's acute criterion (one-hour average) for zinc in freshwater to protect aquatic life is about 120 µg/L, and the chronic criterion (4-day average) is about 120 µg/L for freshwater and 89 µg/L for saltwater. This predicts that high levels of zinc can be toxic to fish. Deficiency causes lens cataracts in fish.

Boron (B)

Boron serves as a vital element for the growth and development of plants and algae. Yet, its significance for humans and animals remains unclear, as the specific biological roles necessary for its essentiality have not been pinpointed. Boron's essentiality has been conclusively established only in zebrafish, which is required for their entire life cycle. Additionally, boron has been found to promote embryonic growth in both trout and zebrafish. Boron has beneficial effects on such functions as reproduction and development, calcium metabolism, bone formation, brain function, insulin and energy substrate metabolism, immunity and the function of vitamin D and steroid hormones (Nielsen, 2014). Compared to fresh water , the boron concentration in the marine environment is relatively high (0.4 mM), efficiently taken up by algae. The harmful effects of compound B on young Chinook and Coho salmon have been documented due to its acute toxicity. According to research, adding boron to their diets has been linked to increased testosterone and estradiol levels (Naghii *et al.* 2011). Insufficient boron intake could potentially lead to a higher risk of osteoporosis.

Iron

Iron, as a crucial trace element, has garnered extensive research attention and is ubiquitous within the cellular composition of vertebrates (Hardy and Kaushik, 2021). Its indispensability extends to numerous biochemical functions, encompassing electron transfer reactions, genetic regulation, oxygen binding and transportation, and cellular proliferation and specialization regulation. Predominantly, iron manifests in vital compounds such as heme proteins, including haemoglobin, myoglobin, and cytochromes. Additionally, non-heme iron-containing enzymes, like iron-sulfur cluster proteins (e.g., NADH dehydrogenase, succinate dehydrogenase, xanthine oxidase), play pivotal roles in energy metabolism. Another subset of iron-utilizing enzymes, such as hydrogen peroxidases, are renowned for their activity on reactive oxygen by-products. The metabolic journey of iron encompasses absorption from the gastrointestinal tract, intracellular transportation, incorporation into proteins, and eventual storage within ferritin across diverse tissues. Despite the significance of fish gills in iron uptake, the process remains relatively inefficient (Copper and Burry, 2007).

Particularly noteworthy is the role of fish gills in iron acquisition during early developmental stages, when digestive functionality is limited (Bury and Grosell, 2003). A comprehensive meta-analysis of various fish species' iron requirements revealed a broad spectrum, ranging from 58.8–166.4 mg kg^{-1}, with substantial variability across all evaluated parameters. Reported iron requirements for specific fish species typically span from 30–170 mg kg^{-1} in their diet, with gibel carp demonstrating an optimal requirement value of 202 mg kg^{-1}. The optimal dietary iron requirement for common carp is typically around 40-80 mg/kg of the diet. Fish with iron deficiency may exhibit changes in behaviour, such as reduced activity levels and altered feeding patterns. Iron deficiency in fish can lead to "iron-deficiency anaemia."Anaemia occurs when there is a decreased number of red blood cells or a decrease in haemoglobin levels in the blood. This can lead to reduced oxygen-carrying capacity, resulting in lethargy, weakness, and a decreased ability to perform everyday activities. In catfish, iron deficiency suppressed hematocrit, haemoglobin, and plasma iron levels and resulted in decreased transferrin saturation.

Iodine

Iodine is crucial for producing thyroid hormones T_3 (triiodothyronine) and T_4 (thyroxine), which regulate cellular activity and growth throughout the body. The metabolism of these hormones and iodine, primarily found as inorganic iodide, are closely intertwined. Thyroid hormones are vital for various bodily functions, including cellular oxidation, blood production, circulation, reproductive health, neuromuscular coordination, and the metabolism of critical nutrients. T_3 is the primary hormone secreted by the thyroid gland and is a potent precursor to T_4. In certain fish species, T_3 exhibits more excellent biological activity than T_4.

Manganese

Manganese plays a crucial role in various physiological processes, including protein and energy metabolism, bone development, synthesis of glycosaminoglycans, cellular defence against oxidative stress, and metabolic regulation. Its significance stems from its dual role as an activator of enzymes and a constituent of metalloenzymes. These enzymes, such as oxidoreductases, lyases, ligases, hydrolases, kinases, and decarboxylases, are vital for numerous biochemical pathways. Manganese-containing metalloenzymes like arginase, pyruvate carboxylase, and Mn superoxide dismutase (MnSOD) are particularly noteworthy. In fish, there's limited understanding regarding the physiological processes involved in manganese uptake through gills and intestines and its subsequent metabolism. However, in mammals, the gastrointestinal absorption and biliary elimination of manganese are vital regulatory mechanisms for maintaining manganese homeostasis, which can be influenced by dietary manganese intake. Manganese absorption competes with iron absorption at absorption sites. Once absorbed, divalent manganese is swiftly cleared by the liver. Adaptive mechanisms such as reduced intestinal absorption, increased liver metabolism, and enhanced biliary excretion come into play during high dietary manganese intake (Raja *et al.*, 2011).

The manganese requirements of fish typically range from 2.5 to 25 mg per kilogram of diet. Furthermore, studies indicate that body and vertebral zinc levels can be reliable indicators of manganese requirements. A meta-analysis of manganese requirements across various fish species suggested approximate values of 10.7, 13.4, and 18.4 mg per kilogram for weight gain, whole-body manganese content, and vertebral manganese content, respectively (Prabhu *et al.*, 2016).

Cobalt

Cobalt, an essential element, is crucial in forming vitamin B12, also known as cobalamin. The active forms of this vitamin are methylcobalamin and 5-deoxyadenosylcobalamin, although hydroxycobalamin and cyanocobalamin, two other forms, can be converted into these active forms. In certain organisms like ruminants' gut microbiota and algae, cobalt sources are utilized for synthesizing vitamin B12. However, monogastric animals and fish rely on dietary intake of vitamin B12 since they cannot produce sufficient amounts through microbial synthesis. Some warm water fish, such as Nile tilapia and hybrid tilapia, can fulfil their vitamin B12 requirements through intestinal microbial synthesis, while others, like channel catfish, cannot. Recommended cobalt levels in the diet vary, with common carp requiring 0.05 to 0.1 ppm. For *Tilapia zillii*, a reported dietary requirement is about 100 mg per kg of diet. Lower concentrations of Cobalt, such as 10 mg per kg of diet, can stimulate gastrointestinal bacterial synthesis of vitamin B12 in fish like Malabar grouper, meeting their dietary needs (Lin *et al.*, 2010). Cobalt deficiency can lead to vitamin B12 deficiency, resulting in various animal health issues, including neurological problems, anaemia and growth impairment. While these effects may also occur in fish, the specific diseases or conditions directly caused by cobalt deficiency have not been extensively studied or well-documented.

Magnesium(Mg)

Magnesium is essential for fish growth, health, and various physiological processes. Fish feeds usually contain magnesium levels ranging from 0.2% to 0.6% of the diet on a dry weight basis. However, these values can change based on the type of fish being cultured, their life stage, water conditions, and other nutritional considerations. Research findings suggest that maintaining a minimum magnesium level of 0.6 grams per kilogram (g kg^{-1}) is necessary to boost magnesium content in both plasma and bones while decreasing overall calcium concentration, which can lead to symptoms of hypercalcinoses. Increasing dietary magnesium up to 3.2 g kg^{-1} showed minimal improvement in fish growth rate. However, the deposition rate of dietary magnesium decreased significantly to 7.4% and 10.7% in fish fed with low- and high-protein diets, respectively (Dabrowska *et al.*, 1991).

It plays a crucial role in various physiological processes within fish, such as enzyme function, energy metabolism, osmoregulation, and bone development. Fish need to maintain a proper balance of ions, including magnesium, in their bodies to regulate their internal osmotic pressure. Proper osmoregulation is crucial for maintaining bodily functions, growth, and overall health. Adequate magnesium levels can contribute to effective osmoregulation and reduce stress on fish, allowing them to allocate more energy to growth. Magnesium is a cofactor for many enzymes—one of the notable enzymes that often require magnesium as a cofactor is DNA polymerase. DNA polymerase is an enzyme responsible for catalyzing the synthesis of DNA molecules during DNA replication, repair, and other DNA-related processes (Hartwig, 2011). Magnesium deficiency or the displacement of Mg^{2+} by other toxic divalent metal ions leads to increased genomic instability, as evidenced by inhibited DNA repair, oxidative stress, ageing, and carcinogenicity (Rosanna and Salvatore, 2012).

Regarding magnesium's intracellular distribution, major intracellular compartments are mitochondria, the nucleus, and the endoplasmatic reticulum. Magnesium assists in the binding of nucleotides and the overall stability of the enzyme-substrate complex during DNA synthesis. Magnesium is closely related to calcium metabolism in fish. Proper calcium levels are vital for bone and scale development. Magnesium helps regulate calcium levels in the fish's body, and any imbalance could affect bone formation and growth. Magnesium has been suggested to have a calming effect on fish, potentially reducing stress caused by various environmental factors.

Potassium (K)

Potassium is one of the constant minerals essential for maintaining normal physiological activities and growth in fish and shrimp. It mainly exists in the fluid inside the cell and plays a critical role in maintaining the cell's acid-base balance, ion balance, water balance and osmotic pressure. It also plays a vital role in protein synthesis, carbohydrate metabolism, phosphorus metabolic reactions, and nerve and muscle cells in the electricity conduction and the transmission of nerve impulses. When fish and shrimp lack potassium, there will be symptoms of anorexia, convulsions, seizures, death, stunted growth, and protein utilization, whereas an excess of potassium intake also depresses animal growth. The appropriate dietary potassium contents in crustaceans are 9 g kg^{-1} for Penaeus japonicus (Roy *et al.*, 2007).

Sodium (Na)

Sodium (Na) maintains cell balance and supports bodily functions. The recommended daily intake falls between 200 and 500 mg (Farquhar *et al.*, 2015). Yet, surpassing the World Health Organization's guidelines of 5 g NaCl/day or 2 g sodium/day for adults may lead to various health issues, including hypertension, heart diseases, gastric cancers, obesity, and an estimated global death toll of 2.5 million annually (WHO, 2013). Most sodium intake originates from processed foods (Nitipong *et al.*, 2020). Salt plays functional roles in enhancing texture, facilitating protein solubility, activating protein extraction, and improving hydration and water retention. Additionally, it alters protein structure, enhancing its cohesion and aiding in moisture and fat retention (Elias *et al.*, 2019).

Calcium (Ca)

Calcium is one of the most abundant cations in the body of a fish. It is closely related to the development and maintenance of the skeletal system. It participates in several physiological processes, including maintaining acid-base equilibrium, osmoregulation, muscle contraction, bone mineralization, blood clot formation, nerve transmission, maintenance of cell membrane integrity, and activation of several necessary enzymes (Lall, 2002). Standard carp diets often contain magnesium levels ranging from 0.1% to 0.5% of the diet. For freshwater fish, a standard guideline for the Ca:P ratio is around 1.6:1 to 2:1 to 2:1. There should be about 1.2 to 2 times more calcium in the diet than phosphorus (Persson *et al.*, 1998). However, this can vary depending on the specific needs of different fish species. In general, most fish meet a significant portion of their calcium needs by absorbing it through their freshwater gills or drinking in saline environments. Carp, red sea bream, striped bass, tilapia, catfish, and chum salmon typically require a low dietary calcium concentration of 0.34% or less. Catfish and tilapia raised in water with low calcium levels need diets containing 0.45% and 0.7% calcium, respectively. Farmed marine fish have a relatively modest calcium requirement ranging from 0.1% to 0.25%. Marine fish may have different requirements due to their natural diet composition and seawater's higher mineral content. Some aquatic species might require a Ca:P ratio closer to 1:1. In skeletal tissue, calcium (Ca) and phosphorus (P) combine to form tricalcium phosphate ($Ca_3(PO_4)^2$), which subsequently transforms into hydroxyapatite through crystalline modifications. Hydroxyapatite is laid down within the organic matrix during the mineralization process.

Phosphorous (P)

The phosphorus (P) requirements, measured in grams per 100 grams, vary among different fish species raised in freshwater. For instance, common carp typically require 0.6 to 0.7 grams per 100 grams of feed, while milkfish demand approximately 0.85 grams per 100 grams (Borlongan and Satoh, 2001). Blue tilapia necessitates around 0.5 grams, Nile tilapia ranges between 0.65 and 0.86 grams, and gibel carp requires between 0.67 and 1.0 grams. Additionally, grass carp typically requires 0.85 grams per 100 grams of feed. Numerous studies investigating the phosphorus requirements of various fish species have revealed adverse effects on weight gain, feed utilization, and bone mineralization when diets are deficient in phosphorus. Exceptions to this pattern include gibel carp (Xie *et al.*, 2017), red drum, gilthead bream (Rodrigues *et al.*, 2001), and European seabass. Fish subjected to mineral-deficient or low-phosphorus diets tend to mobilize minerals like calcium and phosphorus to maintain normal physiological functions. Fish scales emerge as susceptible indicators of phosphorus deficiency, especially in young fish (NRC, 2011). The underlying causes of skeletal deformities induced by low dietary phosphorus in salmonids remain a subject of investigation. Joint, skeletal deformities with low intake of P include curved spines and soft bones in Atlantic salmon, cephalic deformities in the frontal bones of common carp and compressed vertebral bodies resulting in scoliosis in haddock and halibut. The availability of phosphorus to fish varies significantly among different feed ingredients and supplements, as well as due to various dietary factors such as chemical composition, diet digestibility, particle size, nutrient interactions, feed processing methods, and water quality. In fish meal, phosphorus exists predominantly in the form of an insoluble calcium-phosphorus complex called hydroxyapatite, which varies depending on the fish source used and exhibits different bioavailability among fish species. Acid hydrolysis of fish bone by-products enhances phosphorus availability from hydroxyapatite in bones. However, in cereal grains and oilseed meals, phosphorus primarily exists in an organic complex known as phytic acid or phytate, which is largely inaccessible to fish due to the lack of sufficient extracellular phytase in their intestinal tract, unlike in other non-ruminant vertebrates. Phytate negatively impacts protein and amino acid digestibility and utilization efficiency in fish and other animals, with its interactions with proteins influenced by pH levels.

Conclusion

Signs of mineral deficiency in fish encompass various symptoms, including diminished bone strength, loss of appetite, the development of lens cloudiness (linked to zinc deficiency), abnormal skeletal growth (associated with deficiencies in phosphorus, magnesium, and zinc), deterioration of fins (attributed to copper and zinc deficiencies), kidney calcification (resulting from magnesium deficiency or selenium overexposure), enlargement of the thyroid gland (due to iodine deficiency), muscle weakness (caused by selenium deficiency), and a type of anaemia characterized by reduced red blood cell size and colouration (related to iron deficiency). Conversely, an overabundance of minerals can lead to toxicity, whether acquired through diet or gill absorption. Hence, maintaining a delicate equilibrium between mineral insufficiency and excess is crucial for aquatic organisms to uphold internal stability, achieved through mechanisms such as heightened absorption or elimination. Furthermore, the discharge of minerals from uneaten feed or waste and urinary excretion pose a risk of water eutrophication, necessitating careful consideration of feed composition. Attention should also be directed towards accumulating major and trace minerals in Recirculating Aquaculture Systems

(RAS), as reducing the energy expenditure fish allocate to actively obtaining minerals from the water could be advantageous. Investigating how fish regulate mineral intake from water and diet post-feeding initiation could yield valuable insights. Alternatively, exploring the thresholds at which macro and trace mineral concentrations become toxic, potentially disrupting standard cellular ion transport mechanisms, presents another intriguing avenue for research.

References

Antony, Jesu, Prabhu, P., Schrama, J. W. and Kaushik, S. J. (2016). Mineral requirements of fish: a systematic review. *Reviews in Aquaculture, 8*(2), 172-219

Borlongan, I. G. and Satoh, S. J. A. R. (2001). Dietary phosphorus requirement of juvenile milkfish, *Chanos chanos* (Forsskal). *Aquaculture Research, 32*, 26-32.

Boyd, C. E. (2019). *Water quality: an introduction. Springer Nature.*

Boyd, C. E., Tucker, C. S. and Somridhivej, B. (2016). Alkalinity and hardness: critical but elusive concepts in aquaculture. *Journal of the World Aquaculture Society, 47*(1), 6-41.

Bury, N. and Grosell, M. (2003). Iron acquisition by teleost fish. *Comparative Biochemistry and Physiology Part C: Toxicology and Pharmacology, 135*(2), 97-105.

Chanda, S., Paul, B. N., Ghosh, K. and Giri, S. S. (2015). Dietary essentiality of trace minerals in aquaculture-A Review. *Agricultural Reviews, 36*(2), 100-112.

Chavez-Sanchez, C., Martinez-Palacios, C. A., Martinez-Perez, G. and Ross, L. G. (2000). Phosphorus and calcium requirements in the diet of the American cichlid *Cichlasoma urophthalmus* (Gunther). *Aquaculture Nutrition, 6*(1), 1-10.

Colt, J., 2006. Water quality requirements for reuse systems. *Aquaculture* engineering. 34, 143–156

Dabrowska, H., Meyer-Burgdorff, K. H. and Gunther, K. D. (1991). Magnesium status in freshwater fish, common carp (*Cyprinus carpio*, L.) and the dietary protein-magnesium interaction. *Fish Physiology and Biochemistry, 9*, 165-172.

Dawood, M. A., Alagawany, M. and Sewilam, H. (2021). The role of zinc microelement in aquaculture: A review. *Biological trace element research*, 1-13.

Di Rosanna, P., and Salvatore, C. (2012). Reactive oxygen species, inflammation, and lung diseases. *Current pharmaceutical design, 18*(26), 3889-3900.

Elias, M., Laranjo, M., Agulheiro-Santos, A. C. and Potes, M. E. (2020). The role of salt on food and human health. *Salt in the Earth, 19.*

Evans, D. H., Piermarini, P. M. and Choe, K. P. (2005). The multifunctional fish gill: dominant site of gas exchange, osmoregulation, acid-base regulation, and excretion of nitrogenous waste. *Physiological reviews, 85*(1), 97-177.

Fairchild, J.F., Allert, A.L., Sappington, L.C. and Waddell, B., (2005). Chronic toxicity of un-ionized ammonia to early life stages of endangered Colorado pikeminnow

(*Ptychocheilus lucius*) and razorback sucker (*Xyrauchen texanus*) compared to the surrogate fathead minnow (*Pimephales promelas*). *Environmental Contamination and Toxicology*.49, 378–384.

FAO (2020). The state of world fisheries and aquaculture FAO, United Nations.

Farquhar, W. B., Edwards, D. G., Jurkovitz, C. T. and Weintraub, W. S. (2015). Dietary sodium and health: more than just blood pressure. *Journal of the American College of Cardiology, 65*(10), 1042-1050.

Glover, C. N., Bury, N. R. and Hogstrand, C. (2004). Intestinal zinc uptake in freshwater rainbow trout: evidence for apical pathways associated with potassium efflux and modified by calcium. *Biochimica et Biophysica Acta (BBA)-Biomembranes, 1663*(1-2), 214-221.

Hardy, R. W., and Kaushik, S. J. (Eds.). (2021). *Fish nutrition*. Academic press.

Hartwig, A. (2001). Role of magnesium in genomic stability. *Mutation Research/Fundamental and Molecular Mechanisms of Mutagenesis, 475*(1-2), 113-121.

Hassaan, M. S., El-Sayed, A. M. I., Mohammady, E. Y., Zaki, M. A., Elkhyat, M. M., Jarmołowicz, S. and El-Haroun, E. R. (2021). Eubiotic effect of a dietary potassium diformate (KDF) and probiotic (*Lactobacillus acidophilus*) on growth, hemato-biochemical indices, antioxidant status and intestinal functional topography of cultured Nile tilapia (*Oreochromis niloticus*) fed diet free fishmeal. *Aquaculture, 533*, 736147.

He, Y., Zhao, T., Lv, W. H., Tan, X. Y., Song, Y. F., Song, C. C. and Luo, Z. (2023). Interactive effects of dietary zinc and water temperature on growth performance, Zn transport, heat shock response and lipid metabolism in the livers of juvenile yellow catfish *Pelteobagrus fulvidraco. Aquaculture, 562*, 738795.

John P. R, ed. (2011). *Minerals*; p. 1. In the series *Geology: Landforms, Minerals, and Rocks*. Rosen Publishing Group. ISBN 978-1615304899

Katya, K., Lee, S., Yun, H., Dagoberto, S., Browdy, C. L., Vazquez-Anon, M. and Bai, S. C. (2016). Efficacy of inorganic and chelated trace minerals (Cu, Zn and Mn) premix sources in Pacific white shrimp, *Litopenaeus vannamei* (Boone) fed plant protein based diets. *Aquaculture, 459*, 117-123.

Kucukbay, Z., Yazlak, H., Sahin, N., Tuzcu, M., Cakmak, M. N., Gurdogan, F. and Sahin, K. (2006). Zinc picolinate supplementation decreases oxidative stress in rainbow trout (*Oncorhynchus mykiss*). *Aquaculture, 257*(1-4), 465-469.

Lall, S. P. (2002). The minerals. Fish nutrition. *Elsevier Academic Press, San Diego, CA, USA*, 259-308.

Lin, Y. H., Wu, J. Y. and Shiau, S. Y. (2010). Dietary Cobalt can promote gastrointestinal bacterial production of vitamin B12 in sufficient amounts to supply the growth requirements of grouper, *Epinephelus malabaricus*. *Aquaculture, 302*(1-2), 89-93.

Loitzenbauer, E., and Mendes, C. A. B. (2012). Salinity dynamics as a tool for water resources management in coastal zones: An application in the Tramandai River basin, southern Brazil. *Ocean and coastal management*, *55*, 52-62.

Martins, C. I., Pistrin, M. G., Ende, S. S., Eding, E. H. and Verreth, J. A. (2009). The accumulation of substances in Recirculating Aquaculture Systems (RAS) affects embryonic and larval development in common carp *Cyprinus carpio*. *Aquaculture*, *291*(1-2), 65-73.

McGurk, M.D., Landry, F., Tang, A. and Hanks, C.C. (2006). Acute and chronic toxicity of nitrate to early life stages of lake trout (*Salvelinus namaycush*) and lake whitefish (*Coregonus clupeaformis*). *Environmental toxicology and chemistry*. 25, 2187–2196

Naghii, M. R., Mofid, M., Asgari, A. R., Hedayati, M., and Daneshpour, M. S. (2011). Comparative effects of daily and weekly boron supplementation on plasma steroid hormones and proinflammatory cytokines. *Journal of trace elements in medicine and biology*, *25*(1), 54-58.

National Research Council (2011). *Nutrient requirements of fish and shrimp*. National Academic Press.

Nielsen, F. H. (2014). Update on human health effects of boron. *Journal of Trace Elements in Medicine and Biology*, *28*(4), 383-387

Nitipong, J., Kamonwan, R. and Teeraporn, K. (2020). Effects of low sodium chloride substitutes on physico-chemical and sensory properties of Kapi, a fermented shrimp paste, during fermentation. *Journal of Microbiology, Biotechnology and Food Sciences*, *9*(4).

Persson, P., Sundell, K., Björnsson, B. T., and Lundqvist, H. (1998). Calcium metabolism and osmoregulation during sexual maturation of river running Atlantic salmon. *Journal of Fish Biology*, *52*(2), 334-349.

Philip, A. J. P., Fjelldal, P. G., Remo, S., Silva, M., Holme, M. H., Lock, E. J. and Waagbo, R. (2023). Dietary zinc, selenium and water temperature during early seawater phase influences the development of vertebral deformities and cataract in adult Atlantic salmon. *Aquaculture*, *572*, 739529.

Pimentel Rodrigues, A. M. and Oliva Teles, A. (2001). Phosphorus requirements of gilthead sea bream (*Sparus aurata* L.) juveniles. *Aquaculture Research*, *32*, 157-161.

Raja, M. M. M., Raja, A., Imran, M. M., Santha, A. M. I. and Devasena, K. (2011). Enzymes application in diagnostic prospects. *Biotechnology*, *10*(1), 51-59.

Roy, L. A., Davis, D. A., Saoud, I. P. and Henry, R. P. (2007). Effects of varying levels of aqueous potassium and magnesium on survival, growth, and respiration of the Pacific white shrimp, *Litopenaeus vannamei*, reared in low salinity waters. *Aquaculture*, *262*(2-4), 461-469.

Roy, L. A., Davis, D. A., Saoud, I. P. and Henry, R. P. (2007). Effects of varying levels of aqueous potassium and magnesium on survival, growth and respiration of the Pacific white shrimp, *Litopenaeus vannamei*, reared in low salinity waters. *Aquaculture, 262*(2-4), 461-469.

Saoud, I. P., Davis, D. A. and Rouse, D. B. (2003). Suitability studies of inland well waters for *Litopenaeus vannamei* culture. *Aquaculture, 217*(1-4), 373-383.

Soetan, K. O., Olaiya, C. O. and Oyewole, O. E. (2010). The importance of mineral elements for humans, domestic animals and plants: A review. *African journal of food science*, 4(5), 200-222.

Tacon, A. G. (2020). Trends in global aquaculture and aquafeed production: 2000–2017. Reviews in Fisheries Science and Aquaculture, 28(1), 43-56.

Watanabe, T., Kiron, V. and Satoh, S. (1997). Trace minerals in fish nutrition. *Aquaculture, 151*(1-4), 185-207.

Wei, C. C., Luo, Z., Hogstrand, C., Xu, Y. H., Wu, L. X., Chen, G. H. and Song, Y. F. (2018). Zinc reduces hepatic lipid deposition and activates lipophagy via Zn2+/MTF 1/PPAR and Ca^{2+}/CaMKKβ/AMPK pathways. *The FASEB Journal, 32*(12), 6666-6680.

World Health Organization. (2013). Mapping salt reduction initiatives in the WHO European Region.

Xie, D., Han, D., Zhu, X., Yang, Y., Jin, J., Liu, H. and Xie, S. (2017). Dietary available phosphorus requirement for on growing gibel carp (*Carassius auratus gibelio var*. CAS III). *Aquaculture Nutrition, 23*(5), 1104-1112.

Xu, X. W., Song, C. C., Tan, X. Y., Zhong, C. C. and Luo, Z. (2023). Effects of dietary zinc (Zn) levels on growth performance, nutrient composition, muscle development, antioxidant and inflammatory responses in yellow catfish muscle. *Aquaculture Reports, 31*, 101640.

CHAPTER 9
MOST PREVALENT FISH DISEASE AND THEIR ETIOLOGY

Fish Infectious Diseases

Fish infectious diseases are a significant concern in both aquaculture and wild fish populations, with the potential to cause devastating effects on fish health, economic losses, and ecological balance. Like any other living system, diseases are a significant bottleneck in the progress and sustainability of the aquaculture industry (Van Hai, 2015). These diseases are primarily caused by bacterial and viral agents that infect various fish species, leading to morbidity and mortality.

Infectious agents are microorganisms, such as bacteria and viruses, that can spread from one fish to another directly through contact or indirectly through the water. These pathogens can be highly contagious and rapidly spread within fish populations, resulting in disease outbreaks that may severely affect fish stocks and ecosystems.

Understanding the impact of infectious agents on fish populations is crucial for effective disease management and conservation efforts. Fish are essential in aquatic ecosystems as top predators, prey species, and contributors to nutrient cycling. Consequently, the health and stability of fish populations have far-reaching implications for the overall health and balance of the aquatic ecosystem.

Importance of fish health management in aquaculture and wild populations

Fish health management is paramount in aquaculture and wild fish populations. In aquaculture, where fish are bred, reared, and harvested for commercial purposes, disease outbreaks can lead to significant financial losses for fish farmers and the aquaculture industry. Preventing, diagnosing, and controlling infectious diseases in aquaculture is essential for maintaining the productivity and sustainability of this vital food production sector.

In wild fish populations, infectious diseases can harm biodiversity and the overall health of aquatic ecosystems. Disease outbreaks can lead to declines in fish populations, affecting food webs and potentially leading to imbalances in the ecosystem. Moreover, some infectious diseases can have spillover effects, impacting other aquatic organisms or even affecting human health in cases where zoonotic diseases are involved.

Proactive fish health management strategies, including disease monitoring, rapid diagnosis, and the development of effective preventive measures, are crucial for safeguarding both aquaculture and wild fish populations. Understanding the biology and transmission patterns of infectious agents allows scientists and fishery managers to design appropriate control and mitigation strategies to minimize the impact of diseases on fish populations and ecosystems.

This chapter will explore various infectious bacterial and viral diseases affecting fish. We will delve into the common bacterial and viral pathogens, their modes of transmission, the signs and symptoms of infected fish, and the implications of these diseases on aquaculture and wild fish populations. Additionally, we will discuss the current strategies and best practices for managing and preventing infectious diseases to ensure the health and sustainability of fish populations in diverse aquatic environments.

Bacterial disease

Bacterial diseases in fish are major concerns in aquaculture, impacting fish health and leading to economic losses. Common bacterial diseases include *Aeromoniasis*, caused by *Aeromonas* species, which manifests as hemorrhages, ulcers, and fin rot. *Columnaris* disease, caused by *Flavobacterium columnare*, presents with symptoms such as fin erosion, gill lesions, and skin discoloration. Preventive measures like maintaining water quality, proper nutrition, and timely medical interventions are crucial for managing these infections.

Aeromoniasis: The troublesome aeromonas pathogens

Aeromoniasis is a significant bacterial disease affecting various fish species worldwide. Understanding the complexities of this troublesome disease is crucial for effective disease management and maintaining the health of fish populations. It was recently reported that epizootic ulcerative syndrome (EUS) caused by *Aeromonas sobria* significantly damaged fish farms in parts of southeast Asia, such as Bangladesh and India (Rahman *et al.*, 2002).

Etiology of aeromoniasis

Aeromoniasis is caused by several pathogenic strains of the Aeromonas bacteria, most notably *Aeromonas salmonicida*. However, other species, including *Aeromonas hydrophila*, *Aeromonas caviae*, and *Aeromonas veronii*, have also been implicated in the disease. These bacteria are commonly found in aquatic environments, residing in both freshwater and marine ecosystems. Aeromonas pathogens are versatile and opportunistic, capable of infecting a wide range of fish species, including salmonids, catfish, tilapia, and many others.

Pathogenesis of aeromoniasis

The pathogenesis of Aeromoniasis involves a complex interplay of bacterial virulence factors and host susceptibility. Aeromonas bacteria possess an arsenal of virulence determinants, including extracellular enzymes, toxins, and adhesive structures. These virulence factors allow the bacteria to adhere to fish tissues, evade the host's immune system, and cause tissue damage. Once inside the host, the bacteria can multiply rapidly, leading to systemic infections and septicemia.

Clinical manifestations of aeromoniasis

The clinical manifestations of Aeromoniasis can vary depending on the fish species, age, and environmental conditions. Common clinical signs include skin lesions, fin rot, haemorrhages, and ulcerations. Infected fish may also exhibit signs of lethargy, loss of appetite, and increased mortality rates. In severe cases, the disease can progress to systemic infections, affecting multiple organs and leading to high mortality rates, especially in intensively farmed fish.

Host susceptibility and environmental factors

The susceptibility of fish to Aeromoniasis varies widely among different species and even within the same species due to genetic factors, stress levels, and immune status. Juvenile and stressed fish are more susceptible to infections, while healthy and mature individuals may exhibit higher resistance. Environmental conditions, such as water temperature, pH, and pollution levels, also play a significant role in disease outbreaks. Poor water quality and overcrowding can increase the stress levels of fish, making them more vulnerable to Aeromonas infections.

Diagnosis of aeromoniasis

Early and accurate diagnosis of Aeromoniasis is essential for implementing timely disease control measures. Diagnosis is typically based on clinical signs, histopathological examination of affected tissues, and bacterial isolation and identification. Modern molecular techniques, such as PCR and DNA sequencing, have greatly improved the speed and accuracy of Aeromonas species identification.

Control and prevention of aeromoniasis

Effective control and prevention strategies are essential for managing Aeromoniasis in wild and cultured fish populations. Preventive measures include maintaining optimal water quality, implementing strict biosecurity protocols, and avoiding stress-inducing factors. Vaccination has shown promise in some fish species and can enhance disease resistance. Antibiotic therapy should be used judiciously and following responsible antibiotic use guidelines to prevent the development of antibiotic resistance.

Future perspectives

As research continues, further understanding of the pathogenesis and host-pathogen interactions of Aeromoniasis is expected. Developing novel diagnostic tools and more effective vaccines may revolutionize disease management practices. Additionally, exploring the potential of probiotics and immunostimulants to enhance fish immunity against Aeromonas infections holds promise for the aquaculture industry's sustainability.

Columnaris disease in fish - A deadly flavobacterium infection

Columnaris disease, caused by the bacterium *Flavobacterium columnare* (formerly known as *Flexibacter columnaris*), is a highly contagious and devastating bacterial infection affecting various freshwater and marine fish species. Understanding the intricacies of this deadly infection is essential for effective disease management and preserving fish health in both natural and aquaculture environments. Eradication of columnaris disease from culture settings is unlikely since this bacterium is ubiquitous in freshwater environments.

Etiology of columnaris disease

The Gram-negative bacterium Flavobacterium columnare exclusively causes Columnaris disease. This bacterium is a ubiquitous aquatic microorganism capable of surviving in diverse water conditions. Its ability to exist in free-living and parasitic forms contributes to the disease's widespread distribution in freshwater ecosystems worldwide.

Pathogenesis of columnaris disease

The pathogenesis of Columnaris disease involves complex interactions between the bacterium and the fish host. *F. columnare* produces a range of virulence factors, including extracellular enzymes and toxins, facilitating attachment to fish tissues and subsequent colonization. Once colonized, the bacterium causes tissue destruction, leading to characteristic column-shaped lesions, fin erosion, and skin ulcers. The immune response of the fish plays a crucial role in determining the severity of the infection.

Clinical manifestations of columnaris disease

Clinical signs of Columnaris disease vary among fish species and environmental conditions. Infected fish may display white, grey, or yellowish column-shaped lesions on their skin, gills, or fins, which give the disease its name. As the infection progresses, fish may experience loss of appetite, lethargy, and an increased mortality rate. The disease can spread rapidly in crowded aquaculture settings or stressed fish populations.

Factors influencing disease outbreaks

Several factors influence the occurrence and severity of Columnaris disease outbreaks. Water temperature, pH, and oxygen levels can affect the bacterium's survival and virulence. Stressful conditions, such as overcrowding, poor water quality, and abrupt changes in environmental parameters, can compromise fish's immune function, rendering them more susceptible to infection. Additionally, other pathogens or concurrent diseases can exacerbate the impact of Columnaris disease.

Diagnosis of columnaris disease

Early diagnosis is essential for implementing timely control measures. Diagnosis of Columnaris disease involves observing clinical signs, conducting bacterial isolation and identification, and performing histopathological examinations on infected tissues. Molecular techniques like PCR are valuable tools for accurate species identification and disease confirmation.

Control and prevention of columnaris disease

Effective control and prevention strategies are essential to manage Columnaris disease in wild and cultured fish populations. Prevention measures include maintaining optimal water quality, implementing proper hygiene and biosecurity protocols, and reducing stress in fish. Quarantine measures for newly acquired fish can help prevent the introduction of the bacterium into aquaculture facilities. Vaccination and immunostimulation have shown potential for enhancing fish resistance to Columnaris disease.

Future prospects

Continued research on the epidemiology, virulence mechanisms, and host-pathogen interactions of Columnaris disease is critical to develop more effective prevention and treatment strategies. Further exploration of probiotics and bacteriophages as alternative disease control methods may offer promising avenues for disease management.

Streptococcosis: The pervasive *Streptococcus* pathogens in fish

Streptococcosis is a widespread and economically significant bacterial disease affecting aquatic organisms, including fish. Streptococcal disease in fish was first reported in 1957, affecting cultured rainbow trout in Japan (Hoshina *et al.* 1958). This chapter delves into the etiology, clinical manifestations, impact, and management of streptococcosis in fish populations. Streptococcal pathogens have been identified as a significant threat to wild and farmed fish species, causing substantial economic losses in the aquaculture industry globally.

Streptococcal pathogens in fish

Streptococcus is a diverse genus of Gram-positive bacteria known for their wide range of virulence factors and the ability to cause infections in various hosts, including fish. Streptococcal pathogens primarily belong to the Lancefield groups A, B, C, D, F, and G. Group B Streptococcus (GBS) and *Streptococcus iniae* are among fish's most well-studied species causing streptococcosis.

Etiology and clinical manifestations

Streptococcosis presents a range of clinical signs in fish, including external lesions, exophthalmia (pop-eye), skin ulcerations, fin rot, and lethargy. Internal manifestations may involve septicemia, granulomatous inflammation in internal organs, and ascites. Mortality rates can be high, especially in intensive aquaculture systems. GBS has been associated with streptococcal outbreaks in farmed tilapia, while *Streptococcus iniae* commonly infects many fish species, including rainbow trout, yellowtail, and red drum.

Pathogenesis and virulence Factors

Streptococcal pathogens employ a variety of virulence factors to establish infection. Adhesins, capsules, and various toxins contribute to their ability to colonize host tissues and evade the host's immune response. The Lancefield group-specific carbohydrates on the bacterial surface are pivotal in determining pathogenicity.

Diagnostic methods

Accurate and timely diagnosis of streptococcosis is crucial for disease management. Conventional methods like bacterial culture and biochemical tests remain essential. Still, molecular techniques such as polymerase chain reaction (PCR) and DNA sequencing have gained prominence due to their specificity and rapidity. Immunohistochemistry and serological assays also aid in diagnosing streptococcal infections.

Impact on aquaculture

Streptococcosis has considerable economic implications for the aquaculture industry. Outbreaks can lead to significant fish mortalities, reduced growth rates, and increased feed conversion ratios. The loss of marketable fish and the cost of disease control measures contribute to financial losses for aquaculture producers.

Management and prevention

Effective disease management strategies include using vaccines, antibiotics, and probiotics. Vaccination against streptococcal pathogens has shown promise in reducing disease prevalence and severity. The responsible use of antibiotics and improved biosecurity practices helps minimize disease outbreaks. Probiotics offer a more sustainable approach by promoting beneficial bacterial populations that compete with pathogenic streptococci.

Future directions

Continued research is essential to understand the evolving dynamics of streptococcal infections in fish. This includes investigating emerging strains, studying host-pathogen interactions, and exploring alternative antibiotic treatments. Genomic studies can shed light on the genetic basis of virulence and aid in developing novel therapeutic interventions.

Motile aeromonad septicemia (MAS): A threat to finfish aquaculture

Motile Aeromonad Septicemia (MAS), caused by various species of the Aeromonas, is a significant bacterial disease impacting many finfish species in aquaculture. With its ability to cause devastating outbreaks and economic losses, MAS remains a critical concern for the global aquaculture industry.

Etiology and aeromonas species

Motile aeromonads are Gram-negative bacteria belonging to the genus Aeromonas. Multiple species within this genus, including *A. hydrophila, A. salmonicida, and A. veronii,* have been identified as causative agents of MAS. Aeromonads are ubiquitous in aquatic environments, constantly threatening fish health, particularly in crowded and stress-prone aquaculture settings. A new virulent clonal population of *Aeromonas hydrophila* (vAh) was etiologically determined to be responsible for the MAS outbreaks (Pridgeon *et al.*, 2014).

Clinical manifestations

MAS presents a diverse array of clinical signs depending on the species of Aeromonas involved and the host fish species. Common clinical manifestations include skin ulcerations, haemorrhages, exophthalmia (pop-eye), abdominal swelling, and fin rot. Infected fish may exhibit behavioural changes, such as lethargy, loss of appetite, and increased mortality rates. The severity of symptoms often escalates under stressful conditions.

Pathogenesis and virulence factors

Aeromonads use virulence factors, including adhesins, proteases, cytotoxins, and lipopolysaccharides, to colonize and damage host tissues. Their ability to produce hemolysins and pore-forming toxins contributes to tissue destruction and immune evasion. The interplay between these virulence factors and the host immune response dictates the severity of the disease

Diagnostic methods

Accurate and timely diagnosis of MAS is essential for effective disease management. Conventional methods such as bacterial culture and biochemical tests are commonly used for identification. Molecular techniques like polymerase chain reaction (PCR) and DNA sequencing offer higher specificity and speed in identifying Aeromonas species. Histopathology and serological assays also aid in diagnosing MAS.

Impact on aquaculture

MAS outbreaks can lead to substantial economic losses in finfish aquaculture due to increased mortality rates, decreased growth rates, and additional costs associated with disease control measures. The unpredictable nature of outbreaks underscores the need for robust disease management strategies.

Management and prevention

Managing MAS requires a multifaceted approach that includes improving biosecurity measures, optimizing water quality, implementing vaccination programs, and using antimicrobials judiciously. Vaccination against specific *Aeromonas* species has shown promise in reducing the prevalence and severity of MAS. Water treatment, including disinfection and probiotic supplementation, aids in preventing disease transmission and maintaining a healthy aquatic environment.

Emerging challenges and future directions

As MAS continues to evolve, the emergence of antibiotic-resistant strains and changing environmental conditions pose new challenges. Continued research is necessary to develop effective alternatives to antibiotics, explore host-pathogen interactions, and advance our understanding of the genetic basis of virulence in different Aeromonas species.

Vibrio infections: unravelling the vibrant world of *Vibrio* pathogens

Vibrio infections, caused by various species within the genus *Vibrio*, pose a significant threat to marine and freshwater aquatic organisms, including fish, crustaceans, and molluscs. *Vibrio*sis is a major fish disease among many cultured and wild fish species, leading to significant economic losses. Understanding the dynamics of *Vibrio* infections is crucial for maintaining healthy aquatic ecosystems and sustaining aquaculture practices.

The vibrant genus *Vibrio*

The genus *Vibrio* encompasses many Gram-negative, facultatively anaerobic bacteria that inhabit aquatic environments worldwide. While many *Vibrio* species are harmless, certain strains have evolved to be pathogenic to aquatic organisms and, in some cases, can infect humans upon consumption of contaminated seafood.

Modes of transmission

Vibrio infections can occur through various routes, including direct contact, contaminated water or food consumption, and exposure to carrier organisms. Environmental conditions such as temperature, salinity, and nutrient availability can influence *Vibrio* proliferation and the likelihood of disease outbreaks.

Clinical manifestations

The clinical signs of *Vibrio* infections in aquatic organisms are diverse, reflecting the wide range of *Vibrio* species and host species involved. Common clinical manifestations include ulcerations, haemorrhages, fin rot, tail necrosis, and exophthalmia (pop-eye). Systemic infections may lead to septicemia, ascites, and organ damage. Rapid progression of symptoms often characterizes *Vibrio* infections, particularly under stressful conditions.

Virulence factors and pathogenesis

Vibrio species deploy an arsenal of virulence factors to colonize host tissues, evade immune responses, and cause disease. These factors include adhesins, toxins, secretion systems, and quorum-sensing systems that facilitate coordinated virulence gene expression. The interplay between these factors and host susceptibility dictates the severity of *Vibrio* infections.

Diagnostic approaches

Accurate diagnosis of *Vibrio* infections is essential for effective disease management. Traditional methods like bacterial culture and biochemical tests remain relevant, while molecular techniques such as polymerase chain reaction (PCR) and whole-genome sequencing enhance specificity and speed. Histopathology and immunohistochemistry provide valuable insights into tissue pathology and the distribution of *Vibrio* bacteria.

Impact on aquatic ecosystems and aquaculture:

Vibrio infections can cause substantial mortality and economic losses in aquaculture operations. Outbreaks in wild populations can disrupt ecosystems and lead to population declines. The susceptibility of aquaculture systems to *Vibrio* infections underscores the need for preventive measures and rapid response strategies.

Management strategies

Efforts to manage *Vibrio* infections encompass several vital approaches. Improved biosecurity measures, water quality management, and reduced stress levels contribute to preventing disease outbreaks. Probiotics, competitive exclusion, and prebiotics are emerging strategies to manipulate the aquatic microbiota and mitigate *Vibrio* infections. Additionally, vaccination and developing immune-enhancing feeds are promising avenues for disease control.

Emerging challenges and future directions

The emergence of antibiotic-resistant *Vibrio* strains and the complex interactions between *Vibrio* pathogens, host species, and the environment present ongoing challenges. Research must continue to explore new methods of disease prevention, alternative treatments, and strategies to mitigate the impact of *Vibrio* infections on wild and cultured aquatic organisms.

Viral diseases in fish

Viral diseases significantly threaten both wild and farmed fish populations, leading to severe economic losses and ecological imbalances. This chapter explores two prominent viral diseases: Viral Hemorrhagic Septicemia (VHS) caused by rhabdoviruses and Infectious Pancreatic Necrosis (IPN) caused by birnaviruses. Understanding these viral diseases' biology, transmission, clinical manifestations, and management is crucial for effective disease prevention and control strategies.

Viral Hemorrhagic Septicemia (VHS): A lethal Rhabdovirus affliction

Etiology and Rhabdovirus family

Rhabdoviruses, a family of single-stranded RNA viruses, cause Viral Hemorrhagic Septicemia (VHS). VHS has been recognized as a significant threat to many fish species, including salmonids and cyprinids. Multiple VHS virus genotypes and subtypes have been identified, leading to varying degrees of virulence.

Transmission and clinical manifestations

VHS spreads through waterborne transmission, primarily entering fish through gills. Clinical manifestations include hemorrhagic lesions, abdominal swelling, exophthalmia, and lethargy. The virus disrupts blood vessels, leading to internal and external bleeding. Mortality rates can be high, especially in susceptible populations and stressful conditions.

Diagnosis and impact

Diagnosis involves virus isolation, molecular detection methods (RT-PCR), and serological assays. VHS has a significant impact on both wild fish populations and aquaculture operations. Infected fish may suffer from reduced growth rates, impaired immune responses, and increased susceptibility to secondary infections.

Management and control

Biosecurity measures, quarantine protocols, and movement restrictions are vital for preventing VHS spread. Vaccination strategies and good management practices have shown promise in reducing disease prevalence. Improved water quality and stress reduction also contribute to minimizing VHS outbreaks.

Infectious Pancreatic Necrosis (IPN): Unraveling the Birnavirus threat

Etiology and Birnavirus family

Birnaviruses, a family of double-stranded RNA viruses, cause Infectious Pancreatic Necrosis (IPN). IPN primarily affects salmonids, leading to high mortality rates in both wild and cultured populations. The virus targets the pancreas and other internal organs, causing necrosis and severe organ damage.

Transmission and clinical manifestations

IPN spreads through horizontal transmission, including contact with infected water, contaminated equipment, and carrier fish. Infected fish exhibit various clinical signs, including anorexia, darkened skin, abdominal distension, and exophthalmia. The virus's impact on the pancreas disrupts digestive functions and leads to systemic effects.

Diagnosis and impact

Diagnostic methods for IPN include viral isolation, RT-PCR, and histopathology. IPN has severe economic implications due to its ability to cause mass mortalities and stunted growth in affected fish. Outbreaks in aquaculture operations can lead to significant financial losses.

Management and control

Preventing IPN requires strict biosecurity measures to limit the introduction and spread of the virus. Although challenging due to the variability of Birnavirus strains, vaccination is a promising strategy for disease control. Maintaining optimal water quality, avoiding stressors, and proper nutrition contribute to reducing the susceptibility of fish to IPN.

Viral diseases such as VHS and IPN represent substantial challenges in managing fish populations. Understanding the etiology, transmission, clinical manifestations, and management strategies for these viral diseases is essential for the sustainable health of aquatic ecosystems and aquaculture. Researchers and practitioners can work towards effective disease prevention, control, and mitigation by unravelling the complexities of viral infections.

Infectious Salmon Anemia (ISA): The Orthomyxovirus menace

Infectious Salmon Anemia (ISA) is a highly contagious viral disease that significantly threatens salmonid populations in the wild and aquaculture settings. This section delves into the etiology, transmission, clinical manifestations, economic impact, and strategies for managing ISA.

Unveiling the Orthomyxovirus threat

Viral etiology and taxonomy

An orthomyxovirus, part of the same family as the influenza virus, causes ISA. The virus is officially named the Infectious Salmon Anemia Virus (ISAV). It has distinct genetic variations, leading to various strains with different levels of virulence.

Affected species

Infectious Salmon anaemia primarily affects salmonid fish, including Atlantic salmon (*Salmo salar*), rainbow trout (*Oncorhynchus mykiss*), and coho salmon (*Oncorhynchus kisutch*). This disease has been reported in various regions globally, causing significant economic losses in the aquaculture industry.

Modes of transmission and clinical signs

Transmission routes

ISA spreads primarily through horizontal transmission, involving contact with contaminated water, equipment, or infected fish. The virus can also be vertically transmitted from infected broodstock to their progeny. Stressful conditions, such as poor water quality and handling, increase the susceptibility of fish to infection.

Clinical manifestations

Infected fish may exhibit a range of clinical signs, including lethargy, anaemia, exophthalmia (pop-eye), darkening of the skin, and hemorrhagic lesions. The disease primarily affects the blood and circulatory system, leading to organ damage, particularly in the spleen, kidney, and liver.

Diagnosis and economic impact

Diagnostic methods

Diagnosing ISA involves a combination of virological, serological, and molecular methods. Virus isolation, real-time PCR, and enzyme-linked immunosorbent assays (ELISA) are commonly used for detection. Histopathology and immunohistochemistry help confirm the presence of ISAV in affected tissues.

Economic consequences

ISA outbreaks can have severe economic implications for the aquaculture industry. Infected fish experience reduced growth rates, increased mortality rates, and decreased product quality, resulting in significant financial losses. Trade restrictions may be imposed to prevent disease spread.

Mitigation and control strategies

Biosecurity measures

Strict biosecurity protocols are crucial to preventing ISA spread. These include limiting the movement of fish, disinfection of equipment, and proper waste management. Isolating new stock and testing for ISAV before introduction are vital preventive measures.

Surveillance and early detection

Regular health monitoring and surveillance programs enable early detection of ISA outbreaks. This allows for rapid intervention, containment, and mitigation of the disease's impact.

Vaccination efforts

Vaccination against ISA is a primary strategy for disease control. Several types of ISA vaccines have been developed, including inactivated whole virus vaccines and subunit vaccines. Vaccination reduces the severity of the disease, lowers mortality rates, and minimizes production losses.

Infectious Salmon Anemia remains a formidable challenge in wild and aquaculture environments. The ability of ISAV to rapidly spread and cause significant economic losses emphasizes the importance of proactive disease management strategies. By combining strict biosecurity measures, effective surveillance, vaccination, and international collaboration, stakeholders can work towards reducing the impact of this orthomyxovirus menace on salmonid populations and the aquaculture industry.

Koi Herpesvirus (KHV): An emerging epidemic in ornamental fish

Koi Herpesvirus (KHV) is a significant viral pathogen that has emerged as a major threat to ornamental fish, particularly koi and common carp. A herpes-like virus designated koi herpesvirus (KHV) was first isolated in the USA in 1998 following outbreaks of koi and common carp in Israel and the USA (Hedrick *et al.*, 2000). This section delves into the etiology, transmission, clinical manifestations, economic impact, and management strategies associated with KHV outbreaks.

KHV: A threat to the ornamental fish industry:

Etiology and classification:

Koi Herpesvirus (KHV) belongs to the family Alloherpesviridae. The virus is officially named Cyprinid herpesvirus 3 (CyHV-3). KHV primarily affects common carp (*Cyprinus carpio*) and its ornamental variety, koi carp (*Cyprinus carpio koi*).

Global Distribution and impact

KHV has gained notoriety due to its global distribution and the severe economic losses it causes in the ornamental fish industry. Outbreaks have been reported in various regions, resulting in high mortality rates and significant declines in fish populations.

Transmission dynamics and clinical symptoms

Transmission Routes

KHV spreads through direct contact, waterborne transmission, and contaminated equipment. The virus can persist in the environment, contributing to the rapid spread of infection. Stressful conditions, such as fluctuations in temperature and poor water quality, increase the susceptibility of fish to KHV.

Clinical manifestations

Infected fish often display a range of clinical signs, including lethargy, loss of appetite, white or mucus-covered patches on the skin, and gill necrosis. Mortality rates can be devastating, with outbreaks resulting in mass die-offs within affected populations.

Diagnostic methods and economic impact

Diagnostic techniques

KHV diagnosis involves a combination of molecular methods, including PCR assays, virus isolation, and real-time PCR. Histopathology and electron microscopy can also aid in confirming the presence of KHV in affected tissues.

Economic consequences

The emergence of KHV has had far-reaching consequences for the ornamental fish trade. Losses incurred due to mass mortalities, quarantine measures, and trade restrictions have led to economic instability within the industry.

Prevention and management

Biosecurity measures

Biosecurity protocols, such as quarantine procedures, movement restrictions, and disinfection protocols, are crucial in preventing KHV outbreaks. Preventing the introduction of infected fish to new environments is a vital aspect of disease control.

Research and vaccine development

Efforts are ongoing to develop effective vaccines against KHV. Vaccination of fish with attenuated or recombinant KHV strains shows promise in reducing the severity of the disease and preventing outbreaks.

Public awareness and education

Raising awareness among fishkeepers, hobbyists, and industry professionals about the importance of disease prevention and early detection can aid in minimizing the spread of KHV.

Koi Herpesvirus remains a significant challenge to the ornamental fish industry, threatening both wild populations and the trade of ornamental fish species. The rapid spread and devastating impact of KHV emphasize the need for effective disease management strategies. By implementing rigorous biosecurity measures, pursuing vaccination research, and educating stakeholders, the industry can work towards reducing the impact of this emerging epidemic and safeguarding ornamental fish populations.

Viral Nervous Necrosis (VNN): The Betanodavirus enigma

Viral Nervous Necrosis (VNN) is a viral disease affecting many fish species, primarily in marine and brackish water environments. This chapter delves into the intricacies of VNN, exploring its etiology, transmission, clinical manifestations, diagnostic approaches, economic impact, and management strategies.

Unraveling the Betanodavirus mystery

Betanodaviruses

VNN is caused by betanodaviruses, a family of small, non-enveloped RNA viruses belonging to the Nodaviridae family. These viruses primarily infect fish, leading to central nervous system disorders and often high mortality rates. The disease was first described in 1990 in Japan in hatchery-reared Japanese parrotfish (*Oplegnathus fasciatus*)

Host susceptibility

Many fish species, including marine and brackish water species, are susceptible to betanodavirus infections. Larvae and juveniles are particularly vulnerable, with high mortalities observed in affected populations.

Modes of transmission and clinical manifestations

Transmission routes

VNN spreads horizontally through various routes, including direct contact with infected fish, contaminated water, and equipment. Vertical transmission from broodstock to progeny is also possible.

Neurological symptoms

Infected fish often exhibit neurological symptoms, such as erratic swimming behaviour, circular swimming patterns, loss of balance, and corkscrew swimming. These clinical signs result from the virus's impact on the central nervous system.

Diagnostic approaches and economic impact

Diagnostic methods

Diagnosing VNN involves a combination of virological, molecular, and histopathological techniques. Virus isolation, RT-PCR assays, and immunohistochemistry aid in confirming the presence of betanodaviruses.

Economic consequences

VNN outbreaks can lead to substantial economic losses in aquaculture operations, particularly in fish production's larval and juvenile stages. Mortality rates and reduced growth rates contribute to financial instability.

Mitigation and control strategies

Biosecurity measures

Strict biosecurity protocols, including effective quarantine measures, disinfection of equipment, and controlling movement of fish, are essential to prevent the introduction and spread of VNN.

Vaccination and immune enhancement

Vaccination strategies have been developed to mitigate the impact of VNN. Inactivated vaccines, recombinant protein vaccines, and DNA vaccines show promise in reducing the severity of the disease and preventing outbreaks.

Environmental management

Maintaining optimal water quality and reducing stressors can help minimize the susceptibility of fish to VNN. Healthy and stress-free fish are more resilient to viral infections.

Viral Nervous Necrosis (VNN) presents a complex challenge to the aquaculture industry and wild fish populations. The enigmatic nature of betanodaviruses and their diverse host range underscore the need for proactive disease management. By implementing stringent biosecurity measures, investing in research for effective vaccines, and promoting healthy fish husbandry practices, stakeholders can work towards unravelling the enigma of VNN and ensuring the sustainability of fish populations.

Fungal disease

Fungal diseases in fish are often caused by water molds, particularly species of the genera *Saprolegnia* and *Achlya*. These fungi can infect both freshwater and saltwater fish and are typically opportunistic pathogens, meaning they often infect fish that are already weakened by other stressors such as poor water quality, injury, or other infections.

Common fungal Pathogen

1. ***Saprolegnia* spp.**: The most common water mold affecting fish, causing a condition often referred to as "cotton wool disease" due to the appearance of white, cotton-like growths on the skin and gills.
2. ***Achlya* spp.**: Another genus of water molds that can infect fish, though less commonly than Saprolegnia.

Symptoms

- **White or grayish cotton-like growths** on the skin, gills, or fins.
- **Lethargy:** Infected fish may become less active.
- **Loss of appetite:** Fish may eat less or stop eating altogether.
- **Skin ulcers**: Open sores or lesions on the skin.
- **Gill damage:** Difficulty in breathing or rapid gill movement due to gill infection.

Parasitic disease

Parasitic diseases in fish are caused by various types of parasites, including protozoans, helminths (worms), and crustaceans. These diseases can affect fish health, growth, and survival, and are of significant concern in both wild and aquaculture settings. Here are some common parasitic diseases in fish.

Ichthyophthiriasis (Ich or white Spot Disease)

Ichthyophthiriasis is caused by protozoan parasite *Ichthyophthirius multifiliis*. White spots on the skin, gills, and fins; fish may exhibit rubbing or flashing against objects are the main symptoms of disease. This can be treated by increasing water temperature, formalin, copper sulfate, or potassium permanganate treatments.

Trichodinosis

It is caused by protozoan parasites of the genus *Trichodina*. Excessive mucus production, skin lesions and respiratory distress are the main signs of disease. Formalin or salt baths can be given to treat the symptoms.

Costiasis

This disease is caused by protozoan parasite *Ichthyobodo necator* and the symptoms on fish body includes cloudy skin, increased mucus production, lethargy and weight loss. Fish can be treated by using formalin, copper sulfate, or salt treatments.

Diagnostic techniques for fish infectious diseases

The health of aquatic life, mainly fish, is paramount to the sustainability of natural ecosystems and aquaculture. In fish health management, diagnosing infectious diseases is a critical endeavour. In analyzing fish disease, pathogen detection in a tissue sample is conducted by lethal sampling rather than detecting antibodies that indicate a particular disease. Still, nonlethal sampling is recommended for high-value fishes like ornamental fish. Before the era of sophisticated molecular tools, traditional methods played a pivotal role in identifying and understanding fish pathogens. This chapter delves into the conventional diagnostic techniques of isolating, culturing, and staining microorganisms involved in fish infectious diseases.

Traditional methods: Isolation, cultur and staining techniques

Isolation techniques for fish pathogens

Isolating the causative agents of fish diseases is the first step toward effective diagnosis and treatment. Isolating pathogens from infected fish involves meticulous sampling and selective cultivation. Scientists can target specific areas of infection by collecting samples from affected fish tissues or body fluids. These samples are then streaked onto selective agar media, allowing the isolated pathogens to grow as distinct colonies. This approach aids in the identification and characterization of the responsible microorganisms.

Culture techniques for fish pathogens

Culturing fish pathogens in a laboratory environment is crucial for studying their characteristics, behaviour and treatment susceptibility. Traditional culture methods involve placing pathogenic microorganisms in appropriate growth media, giving them the necessary nutrients to multiply. This process enables researchers to observe these pathogens' growth patterns, colony morphologies, and biochemical properties. By growing pathogens in controlled conditions, scientists can better understand their requirements and devise strategies to combat them.

Staining techniques for fish pathogens

Staining techniques have been invaluable in visualizing fish pathogens and understanding their structures. One such technique, the Gram stain, differentiates different types of bacteria based on their cell wall properties. This information aids in identifying the bacteria causing the infection and choosing the appropriate treatment regimen. Stains like Giemsa and Ziehl-Neelsen also help detect specific microorganisms, such as parasites and acid-fast bacteria, respectively.

Limitations and advancements

While traditional diagnostic techniques have been instrumental, they are not without limitations. Some fish pathogens are challenging to culture in a laboratory setting, and the conditions may not accurately mimic their natural environments. Additionally, specific pathogens might be viable only for a limited time after death, complicating the sampling process. However, these challenges have spurred the development of newer, more accurate diagnostic methods, such as molecular techniques and immunological assays.

Legacy and contemporary relevance

The traditional diagnostic techniques highlighted in this chapter have laid the foundation for contemporary fish health management. While advanced methods like PCR-based assays and next-generation sequencing now dominate the field, the principles of isolation, culture, and staining remain essential for understanding the basic biology of fish pathogens. Furthermore, traditional techniques still find application in resource-limited settings where high-tech methods may not be feasible.

These techniques have provided valuable insights into the world of fish pathogens and paved the way for the development of more advanced and precise diagnostic tools. As we move forward in our quest to safeguard aquatic health, we must acknowledge the importance of these traditional methods that have stood the test of time.

Molecular approaches: PCR, RT-qPCR and Next-generation sequencing

As a critical industry, aquaculture faces constant challenges in maintaining the health of farmed fish populations. Modern techniques have emerged as powerful tools in the pursuit of accurate and timely diagnosis of infectious diseases.

Polymerase chain reaction (PCR) for disease diagnosis

PCR revolutionized the field of diagnostics by allowing the amplification of specific DNA sequences. In aquaculture, PCR enables the detection of pathogenic DNA from small tissue samples or body fluids of infected fish. By targeting unique regions of the pathogen's genome, scientists can rapidly identify the presence of the pathogen, aiding in early disease detection. PCR's sensitivity, speed, and specificity have transformed how fish diseases are diagnosed, offering a valuable tool for aquaculture management. PCR assays can identify about two cells of *Aeromonas salmonicida*, while 16S rRNA sequencing can detect 103 colony-forming units of *A. salmonicida* (Austin 2019).

Real-time quantitative PCR (RT-qPCR) for quantification

RT-qPCR takes PCR a step further by detecting the presence of pathogenic DNA and quantifying it. This technique measures the increase in DNA amplification in real time, providing information about the severity of the infection. RT-qPCR is particularly useful for monitoring disease progression, assessing treatment efficacy and evaluating the overall health of fish populations in aquaculture settings.

Next-generation sequencing (NGS) for pathogen characterization

The advent of NGS has transformed genomics research and disease diagnostics. In aquaculture, NGS allows for comprehensive analysis of entire pathogen genomes, offering insights into their genetic diversity and potential virulence factors. By sequencing the DNA or RNA of pathogens present in fish samples, scientists can identify the specific pathogens and understand their evolutionary relationships and adaptability, crucial for devising effective disease management strategies.

Serological assays: ELISA and immunofluorescence

The health of fish in aquaculture hinges on timely and accurate disease diagnosis. This chapter delves into serological assays, focusing on two fundamental techniques: Enzyme-Linked Immunosorbent Assay (ELISA) and Immunofluorescence.

Enzyme-linked immunosorbent assay (ELISA)

ELISA is a widely used serological technique for detecting and quantifying antibodies or antigens in a sample. In aquaculture, ELISA aids in diagnosing infectious diseases by detecting antibodies produced in response to specific pathogens. Analyzing fish serum or tissue samples provides valuable information about the immune status of the fish population and the prevalence of particular diseases within an aquaculture facility. We recently reported using the ELISA technique to detect infectious pancreatic necrosis virus (IPNV) antigen in infected fish and cell cultures (Dixon and Hill, 1983).

Immunofluorescence for visualizing pathogens

Immunofluorescence is a microscopic technique that employs fluorescently labelled antibodies to detect specific antigens within tissue samples. In aquaculture, this technique allows for visualizing and identifying pathogens within fish tissues. By tagging pathogens with fluorescent markers, researchers can precisely locate and identify the pathogens responsible for infections. Immunofluorescence contributes to our understanding of disease pathology and aids in accurate diagnosis.

As the aquaculture industry grows, so does the need for advanced diagnostic techniques. Molecular approaches like PCR, RT-qPCR, and NGS provide rapid and precise insights into pathogen presence and characteristics. Meanwhile, serological assays like ELISA and Immunofluorescence offer valuable information about immune responses and pathogen visualization. By combining these modern techniques with traditional methods, aquaculture can proactively manage fish health and contribute to sustainable aquatic ecosystems.

Histopathology: The art of microscopic examination in disease diagnosis

Within the intricate world of fish health management in aquaculture, the art of histopathology emerges as a crucial diagnostic technique. Histopathology involves the microscopic examination of tissue samples to uncover the underlying cellular and structural changes caused by infectious diseases. This chapter delves into histopathology and its pivotal role in diagnosing fish contagious diseases.

The essence of histopathology

Histopathology allows scientists to uncover the cellular responses and alterations within fish tissues in response to infections. Scientists can create thin sections that are stained and examined under a microscope by collecting tissue samples from infected fish and subjecting them to specialized processing. These sections reveal changes in cell morphology, tissue architecture, and the presence of pathogens, providing valuable insights into the nature of the disease.

Preparation of tissue samples

The journey of histopathology begins with carefully collecting and preserving tissue samples. Fish tissues affected by infections are carefully dissected, fixed, and embedded in paraffin wax. Thin sections, sliced from these blocks, are placed on glass slides and subjected to various staining techniques. Hematoxylin and eosin (H and E) staining is a common choice, imparting colour to nuclei and cytoplasm, thereby highlighting cellular structures.

Insights gained from histopathology

Histopathological examination reveals a wealth of information about the progression and impact of diseases. Pathological changes such as inflammation, tissue necrosis, fibrosis, and the presence of infectious agents provide critical clues for accurate diagnosis. Moreover, histopathology aids in understanding the mechanism of disease spread, identifying potential reservoirs of infection, and assessing the effectiveness of therapeutic interventions.

Challenges and expertise

Histopathology is not without challenges. The interpretation of tissue sections requires a skilled eye and a deep understanding of fish anatomy and histology. Additionally, some pathogens may not be easily detectable through routine staining, necessitating specialized techniques. However, the wealth of histopathological information makes it an invaluable tool for diagnosing fish infectious diseases.

Integrating histopathology with modern techniques

While molecular and serological techniques offer rapid insights, histopathology provides a level of detail that other methods cannot match. Integrating histopathological findings with molecular data enhances the overall understanding of the disease. For instance, correlating the presence of specific pathogens with the observed tissue changes can lead to more accurate and comprehensive diagnoses.

Histopathology is a cornerstone in aquaculture's arsenal of diagnostic techniques for fish infectious diseases. By meticulously examining tissue samples, scientists unlock a microscopic world that reveals the intricate battle between pathogens and host tissues. By continuing to refine this art, aquaculture can achieve better disease management, healthier fish populations, and a more sustainable future for aquatic ecosystems.

Disease prevention and control strategies in aquaculture

As the aquaculture industry expands to meet the global demand for seafood, disease prevention and control, challenges loom ever more significantly. As "prevention is better than treatment," it is advisable to focus on preventing the occurrence of disease rather than treating it. Aquaculture is more complex than terrestrial animal disease control due to the fish's environment, habitat, and nature. Disease detection and characterization are difficult due to limited observation, rapid transmission, and group behaviour. This chapter explores a comprehensive array of strategies that play pivotal roles in maintaining the health and sustainability of fish populations within aquaculture settings.

Biosecurity measures: The first line of defense

Biosecurity is any management action to prevent the introduction of disease-causing agents to aquaculture facilities. Rigorous measures are taken to avoid the introduction of pathogens into aquaculture facilities. These include controlling farm access, quarantining new stock, and implementing disinfection protocols for equipment and water sources. Aquaculture operations can significantly reduce the likelihood of disease outbreaks by minimizing the risk of pathogen entry.

Genetic selection and breeding programs

Genetic advancements have allowed aquaculture producers to breed fish with enhanced disease resistance selectively. By identifying and breeding individuals with naturally strong immune responses, the susceptibility of the overall population to diseases can be reduced. This approach contributes to the creation of robust fish stocks that are better equipped to withstand various pathogens.

Vaccination campaigns

Vaccination has become a cornerstone of disease prevention in aquaculture. Vaccines are widely accepted because there is no risk of drug resistance development in vaccinated animals, and the minority of unvaccinated animals are protected because of herd immunity. Customized vaccines are developed to target specific pathogens that commonly affect farmed fish. Vaccination campaigns can be applied on a large scale, protecting entire fish populations against prevalent diseases. These immunization efforts not only reduce disease-related mortality but also contribute to the overall welfare of the fish.

Nutritional management

Proper nutrition plays a crucial role in maintaining the immune health of fish. Balanced diets rich in essential nutrients bolster the immune system, making fish more resistant to infections. Nutritional strategies can also include feeding feed with immune-enhancing compounds, such as vitamins and probiotics, to booster the fish's defence mechanisms.

Environmental management

A clean and stable aquatic environment is fundamental to disease prevention. Monitoring water quality parameters, controlling temperature, and maintaining appropriate stocking densities is critical. An optimized environment reduces stress on fish, strengthening their immune responses and making them less susceptible to diseases.

Integrated pest management

In aquaculture, pests and parasites can introduce diseases to fish populations. Integrated Pest Management (IPM) strategies involve using non-chemical and chemical methods to control pests and parasites while minimizing environmental impacts. By managing these vectors, aquaculture facilities can significantly reduce disease transmission.

Collaboration and information sharing

Effective disease prevention and control in aquaculture often involve collaboration among stakeholders, including farmers, researchers, government agencies, and industry associations. Sharing information about disease trends, best practices, and emerging threats enhances the collective ability to respond swiftly and effectively to disease outbreaks.

Ethical considerations

Pursuing disease prevention and control strategies in aquaculture should always be grounded in ethical considerations. Ensuring the welfare of the farmed fish, minimizing environmental impacts, and maintaining sustainable practices are paramount. Balancing disease control with ethical principles leads to a responsible and conscientious approach to aquaculture management.

In conclusion, disease prevention and control strategies form the backbone of a successful and sustainable aquaculture industry. By combining biosecurity measures, genetics, vaccination, proper nutrition, environmental management, and collaboration, aquaculture operations can mitigate the impacts of diseases, safeguard fish health, and contribute to the global demand for high-quality seafood.

Refernces

Austin, B. (2019). Methods for the diagnosis of bacterial fish diseases. *Marine Life Science and Technology, 1*, 41-49.

Dixon, P.F. and Hill, B.J. (1983). Rapid detection of infectious pancreatic necrosis virus (IPNV) by the enzyme-linked immunosorbent assay (ELISA). *J. Gen. Virol.*, 64: 321-330.

Gudding, R. (2014). Vaccination as a preventive measure. *Fish vaccination*, 12-21.

Harikrishnan, R., Balasundaram, C. and Heo, M. S. (2011). Fish health aspects in grouper aquaculture. *Aquaculture*, 320(1-2), 1-21.

Hedrick, R.P., Gilad, O., Yun, S., Spangenberg, J.V., Marty, G.D., Nordhausen, R.W., Kebus, M.J., Bercovier, H. and Eldar, A. (2000). A herpesvirus associated with mass mortality of juvenile and adult koi, a strain of common carp. *Journal of Aquatic Animal Health, 12*, 44–57

Hoshina, T., Sano, T. and Morimoto, Y. 1958. A *Streptococcus* pathogenic to fish. Journal of Tokyo University of Fisheries *44*, 57-68.

Mohd. N., Yazid. S.H.M., Daud. H. M., Amal. M.N.A. and Mohamad N.(2019). Costs of management practices of Asian seabass (*Lates calcarifer* Bloch, 1790) cage culture in Malaysia using a stochastic model that includes uncertainty in mortality. *Aquaculture, 510*, 347-352

Peeler, E. J. and Taylor, N. G. (2011). The application of epidemiology in aquatic animal health-opportunities and challenges. *Veterinary Research, 42*(1), 94.

Phu, T. M., Phuong, N. T., Dung, T. T., Hai, D. M., Son, V. N., Rico, A. and Dalsgaard, A. (2016). An evaluation of fish health management practices and occupational health hazards associated with Pangasius catfish (*Pangasianodon hypophthalmus*) aquaculture in the Mekong Delta, Vietnam. Aquaculture Research, *47*(9), 2778-2794.

Pridgeon, J.W., Zhang, D. and Zhang, L. (2014). Complete genome sequence of the highly virulent *Aeromonas hydrophila* AL09-71 isolated from diseased channel catfish in west Alabama. *Genome Announcements* 2 (3), 00450–514

Rahman, M., P. Colque-Navaro, I. Kuhn, G. Huys, J. and R. Mollby, 2002. Identification and characterization of pathogenic *Aeromonas veronii* biovar sobria associated with epizootic ulcerative syndrome in fish in Bangladesh. *Applied Environmental Microbiology, 68*: 650-655.

Romero, J., Feijoó, C. G. and Navarrete, P. (2012). Antibiotics in aquaculture–use, abuse and alternatives. *Health and environment in aquaculture, 159*, 159-198.

Shoemaker, C. A., Klesius, P. H. and Evans, J. J. (2007). Immunization of eyed channel catfish, *Ictalurus punctatus*, eggs with monovalent *Flavobacterium columnare* vaccine and bivalent *F. columnare* and *Edwardsiella ictaluri* vaccine. *Vaccine, 25*(6), 1126-1131.

Van Hai, N. (2015). The use of medicinal plants as immunostimulants in aquaculture: A review. *Aquaculture, 446*, 88-96.

Yoshikoshi K, Inoue K. 1990. Viral nervous necrosis in hatchery-reared larvae and juveniles of Japanese parrotfish, *Oplegnathus fasciatus* (Temminck and Schlegel). *Journal of Fish Diseases, 13*, 69–77

CHAPTER 10

DISEASE MANAGEMENT IN AQUACULTURE

Introduction

Fish diseases are a significant concern in aquaculture, fisheries, and aquarium keeping, as they can lead to substantial economic losses environmental degradation and impact the health and well-being of aquatic organisms. The study of fish diseases encompasses a wide range of pathogens, including bacteria, viruses, fungi, parasites, and environmental stressors that can compromise fish's immune system. Understanding the causes, symptoms, prevention, and treatment of fish diseases is essential for maintaining sustainable aquatic ecosystems and ensuring the health of fish populations.

Significant diseases in aquaculture - Aquaculture, or farming aquatic organisms such as fish, shellfish, and aquatic plants, is susceptible to various diseases. These diseases can have significant economic and environmental impacts on aquaculture operations. Here are some common and important diseases that affect aquaculture:

1. **Infectious Salmon Anemia** (ISA): ISA is a viral disease primarily affecting salmonids, particularly Atlantic salmon. It causes severe anaemia and can lead to high mortality rates in infected fish.
2. **Viral Hemorrhagic Septicemia** (VHS): VHS is a viral disease affecting many fish species. It causes haemorrhaging, organ damage, and high mortality rates. VHS has been a significant concern in both marine and freshwater aquaculture.
3. **Infectious Pancreatic Necrosis** (IPN): IPN is a viral disease that affects various fish species, including salmonids. It primarily targets the pancreas and can lead to high mortality rates, especially in young fish.
4. **Bacterial Coldwater Disease** (BCWD): BCWD is a bacterial infection primarily affecting salmonids, particularly trout and salmon. It can cause

severe mortality in coldwater environments, leading to significant economic losses.

5. ***Aeromonas* Infections**: Aeromonas bacteria can cause various diseases in fish, including furunculosis and hemorrhagic septicemia. These infections can result in skin lesions, fin rot, and systemic infections.
6. **White spot disease**: White Spot Disease, caused by the white spot syndrome virus (WSSV), is a highly contagious viral infection that affects shrimp and other crustaceans. It leads to white spots on the exoskeleton and high mortality rates.
7. **Early Mortality Syndrome** (EMS): EMS, also known as Acute Hepatopancreatic Necrosis Disease (AHPND), is a bacterial infection that affects shrimp. It causes damage to the hepatopancreas, leading to high mortality rates in affected shrimp populations.
8. **Ichthyophthiriasis** (Ich) is a parasitic infection caused by the protozoan *Ichthyophthirius multifiliis*. It affects a wide range of freshwater and marine fish species, developing white spots, skin irritation, and respiratory problems.
9. **Amoebic Gill Disease** (AGD): AGD is caused by the amoeba Neoparamoeba perurans and primarily affects farmed salmonids. It leads to gill damage respiratory distress, and can result in significant mortality in affected fish.

These are just a few examples of diseases that can impact aquaculture operations. Aquaculturists must maintain strict biosecurity measures, proper nutrition, and vigilant monitoring to prevent and manage disease outbreaks in their facilities.

Vibrosis

- **Causative Agent -**Within the Vibrionaceae, the species causing the most economically severe diseases in marine culture are *Listonella* (*Vibrio*) *anguillarum*, *Vibrio ordalii*, *V. salmonicida* and *V. vulnificus* biotype. (Colquhoun et. al,2014)
- **Symptoms -**Vibriosis generally leads to disseminated systemic infections for which macroscopically visible and non-specific external signs, including necrotic lesions and varying degrees of cutaneous haemorrhage of body and fins, with or without exophthalmia. Internally, extensive inflammation, haemorrhage and ascites can be observed (Colquhoun *et.al*, 2014).
- Extracellular toxins, such as proteases (Denkin and Nelson, 2004; Varina *et al.*, 2008) and hemolysins, undoubtedly play a role in vibriosis, as intramuscular injection of extracellular bacterial products has resulted in the formation of lesions consistent with pathological changes observed during natural infections.
- **Prevention and control -** Vaccination against vibriosis caused by *V. anguillarum* and *V. ordalli* is generally successful, with few outbreaks of

disease registered in vaccinated stocks. Losses to vibriosis are the least common among those species of fish that can be immunized in freshwater before seawater transfer, thus allowing the development of a specific and protective response before challenge with the virulent bacteria. In eels, repeated revaccination may be required to protect *V. vulnificus* until harvest (Esteve-Gassent *et al.*, 2004)

Winter ulcer

- "Winter ulcer" is a disease affecting sea-farmed Atlantic salmon reared at cold temperatures. It occurs mainly during the winter season.
- **Causative agent** - *Moritella viscosa* is the etiological agent of winter ulcer disease, causing considerable mortality and financial loss in salmonids.
- **Symptoms** -The disease is characterized by skin ulcers confined to scale-covered parts of the body surface, and often diffuse or petechial haemorrhage in internal organs is also present (Bruno *et al.*, 1998). Moribund fish become lethargic and swim erratically, have enlarged livers and plasma enzymes suggestive of liver damage, and develop ulcers to the head/jaw.
- **Prevention**: Bacteria can also inhibit growth and kill competitors by secreting antimicrobial compounds toxic to other bacteria. Bacteriocins are proteinaceous molecules that inhibit the growth of similar or closely related bacteria species through lethal membrane potential and integrity disruption.

Photobacteriosis

- **Causative agent** is also known as pasteurellosis and is caused by the halophilic bacterium *Photobacterium damselae* subsp. piscicida (formerly *Pasteurella piscicida*), which was initially isolated from mortalities occurring in natural populations of white perch (*Morone americanus*) and striped bass in 1963 in Chesapeake Bay. Since 1969, pasteurellosis has been one of the most critical diseases in Japan, affecting mainly yellowtail (*S. quinqueradiata*)
- **Symptoms and prevention** - In general, if the fish are vaccinated very early in age, the protection period will be limited, and they will need a revision approximately a month later. In addition, if the disease occurs during the growing period, a third vaccination must be given when fish reach 30–50 g (Romalde *et.al*, 2014). As mentioned before in this chapter, the best protection against photobacteriosis in sea bream and sea bass is obtained with an ECP-enriched bacterin, with a first dip immunization at larval stage (average 0.05 g) and a second vaccination of the fingerlings (1–2 g) in the hatchery, followed by an oral or an injection booster vaccination at the on growing site (Le Breton, 2009).

Furunculosis

- **Causative Agent** -*A. salmonicida* subsp. salmonicida is the causative agent of the so-called "typical" furunculosis, which causes economically devastating losses in cultivated salmonids in fresh and marine waters. (Toranzo *et.al.*, 2005) It also affects a variety of non-salmonid fish and shows a widespread distribution.
- **Symptoms and prevention** - The impact of the "typical" furunculosis may lead to a deep infection of the hair follicle, leading to abscess formation with accumulation of pus and necrotic tissue (Chiller *et al.*,2001). It also affects a variety of non-salmonid fish and shows a widespread distribution.

Marine flexibacteriosis

- *Tenacibaculum maritimum* (formerly *Cytophaga marina, Flexibacter marinus* and *F. maritimus*) is the causative agent of flexibacteriosis in marine fish (Wakabayashi *et al.*, 1986; Bernardet and Grimont, 1989). Several other names, such as "gliding bacterial diseases of sea fish", "eroded mouth syndrome", and "black patch necrosis", were used to designate the disease caused by this bacterium. Marine flexibacteriosis is widely distributed in cultured and wild fish in Europe and Japan.

Pseudomonadiasis

- Among the Pseudomonas species recovered from diseased fish (*P. chlororaphis, P. anguilliseptica, P. fluorescens, P. putida, P. plecoglossicida*), *Pseudomonas anguilliseptica* is considered the most significant pathogen for cultured fish (Toranzo and Barja, 1993, Austin, 2010).
- *P. anguilliseptica* was initially described in 1972 as the etiological agent of "Sekiten-bio" or "red spot disease", which caused massive mortalities in pond-cultured Japanese eel in Japan (Wakabayashi *et al.*, 1986)

Streptococcosis

- Streptococcal infection of fish is considered a reemerging disease affecting various wild and cultured fish worldwide (Romalde and Toranzo, 2002). Classification of Gram-positive cocci based on DNA–DNA hybridization coupled with 16S sequencing has shown that at least five different species are considered of significance as fish pathogens: *Lactococcus garvieae* (syn. *Enterococcus seriolicida*), *Lactococcus piscium*.

Bacterial kidney disease

- **Causative Agent** - Bacterial kidney disease (BKD), caused by the Gram-positive diplobacillus *Renibacterium salmoninarum*, is a chronic systemic disease of salmonids, which causes mortality in cultured fish in fresh and marine environments (Office International des Epizooties, 2000). The pathogen has also been found in wild fish populations.

- **Symptoms -** Symptoms include unilateral or bilateral exophthalmia with ocular lesions. (*Eissa at el.,*2006) with skin discolouration and dark pigmentation, swelling of the abdomen, pale gills (anaemia), anal vent haemorrhage as well as corneal opacity have also been reported. Adult infected fish display belly rash on the skin and muscle cavitation.

Mycobacteriosis (fish tuberculosis)

- Mycobacteriosis in fish (or fish tuberculosis) is a subacute to chronic wasting disease known to affect nearly 200 freshwater and saltwater species. Although *Mycobacterium marinum* is considered the primary causative agent of fish mycobacteriosis, a significant number of Mycobacterium species associated with tubercle granulomas in cultured, aquarium and wild fish populations have been described: *M. marinum, M. fortuitum, M. chelonae, M. smegmatis, M. abscessus, M. neonarum, M. simiae, M. scrofulaceum,*

Piscirickettsiosis

- Piscirickettsiosis is a septicemic condition of salmonids (Office International des Epizooties, 2000). The causative agent of the disease is Piscirickettsia salmonis (Fryer *et al.*, 1992), a non-motile Gram-negative, obligately intracellular bacterium. The disease was described for the first time in 1989, affecting coho salmon cultured in Chile (Bravo and Campos, 1989, Branson *et.al*, 1991)

Schistosomiasis

- Schistosomiasis is endemic in Africa, South America and the Far East. Of the infected subjects, 60% are symptomatic, and 10% had renal dysfunction (Verjee *et al.*, 2019). The lower urinary tract is the primary site of *S. haematobium* infection. Granulomatous response around its ova produces small "pseudotubercles" in the bladder mucosa, which consolidate to form sessile masses and ulcerate. Presentation is usually with microscopic to macroscopic hematuria. Progressive disease leads to fibrosis and bladder calcification.

  The possibility of diseases increases in fishes when fish are reared at higher densities than naturally, and the contact between them increases. Another reason for the occurrence of fish pathology is when infected fish are not removed as promptly from the farm as they would be by natural predators. Farmed fish are more closely and easily observed than wild fish. Thus, pathogens that generally exist in low numbers and do not cause disease in the wild may result in disease in fish.

  When finfish aquaculture operations are in the marine environment, water moves freely between farms and the ocean. This may include the amplification and transmission of disease from farmed to wild fish, and the introduction of pathogens and parasites occurs when fish are transported.

- **Identification** – To identify fish pathogens, a variety of molecular techniques, including amplification of nucleic acids, polymerase chain reaction (PCR), nested PCR, real-time PCR, multiplex PCR, loop-mediated isothermal amplification (LAMP), DNA microarrays, and nucleotide sequencing assays, that are commonly used. (Abdelsalam *et.al*, 2023).
- **Control and management** -For the better growth of aquaculture, it is essential to practice more and more management methods. These include vaccines, probiotics, limiting culture density, high-quality diets and, when appropriate, judicious use of antibiotics. Out of all these practices, Antibiotics are considered a method of last resort, and their use has declined drastically both in the United States and worldwide.

Following farm sites and pest management strategies, therapeutics are used to control farm parasites. Cleaner wrasses and other species have also been used commercially to reduce parasites.

Infectious Salmon Anemia (ISA)

Infectious Salmon Anemia (ISA) is a viral disease that affects salmonid fish, particularly Atlantic salmon (Salmo salar). It is caused by the infectious salmon anaemia virus (ISAV), which belongs to the Orthomyxoviridae family (Lyngstad *et. al.*,2018). ISA primarily affects farmed salmon populations but can also impact wild salmon.

- **Transmission:** ISA can spread through direct contact between infected and susceptible fish or contaminated water, equipment, or fish feed. It can also be transmitted vertically from infected broodstock to their offspring.
- **Symptoms:** Infected fish may exhibit various clinical signs, including anemia, hemorrhaging, lethargy, and swimming abnormalities. Other symptoms may include pale gills, fluid accumulation in the abdomen, and a distended or dark spleen.
- **Impact:** ISA can cause significant economic losses in affected aquaculture operations. Infected fish often experience high mortality rates, leading to reduced production and financial losses for fish farmers.
- **Diagnosis:** Laboratory testing, including molecular techniques like polymerase chain reaction (PCR), is used to confirm the presence of ISAV in infected fish. Blood samples, tissues, or gill swabs are collected for analysis. (Moreira *et.al*, 2021).
- **Prevention and control:** Prevention and control measures for ISA include strict biosecurity protocols, regular health monitoring of fish populations, and implementing measures to minimize the introduction and spread of the virus. Vaccination is also used in some regions. It's worth noting that the specific regulations, surveillance programs, and management practices related to ISA can vary between countries and territories. Suppose you

suspect the presence of Infectious Salmon Anemia in a specific aquaculture context. In that case, it is recommended to consult local authorities, fish health professionals, or relevant regulatory bodies for specific guidance and information.

Viral Hemorrhagic Septicemia (VHS)

Viral Hemorrhagic Septicemia (VHS) is a highly contagious viral disease affecting a wide range of fish species in freshwater and marine environments. It is caused by the viral hemorrhagic septicemia virus (VHSV), which belongs to the Rhabdoviridae family (Vinay *et.al.*, 2013).

- **Transmission:** VHS can be transmitted through direct contact between infected and susceptible fish or waterborne routes. It can also be spread through contaminated equipment, handling practices, and infected fish eggs.
- **Symptoms**: Infected fish may exhibit various symptoms, including haemorrhaging (bleeding), bloated or protruding eyes, darkening of the skin, swollen abdomen, and erratic swimming behaviour. Internal haemorrhaging and organ damage can also occur (Isidan *et.al.*,2011).
- **Impact:** VHS can cause high mortality rates in affected fish populations, leading to significant economic losses for aquaculture operations. It can also impact wild fish populations, including freshwater and marine species.
- **Diagnosis:** Laboratory testing is used to confirm VHS infection in fish. Techniques such as virus isolation, PCR and serological assays are employed to detect the presence of VHSV in fish tissues or blood samples.
- **Prevention and control:** Measures to prevent and control VHS include strict biosecurity protocols, disinfection of equipment and facilities, appropriate handling and disposal of infected fish, and implementing surveillance and monitoring programs. Vaccination is also used in some cases.

  It's important to note that VHS is considered a reportable disease in many countries, and specific regulations and control measures may be in place. Suppose you suspect the presence of Viral Hemorrhagic Septicemia in an aquaculture setting or observe suspicious fish mortality. In that case, contacting relevant fish health authorities or regulatory bodies for guidance and assistance is essential.

Infectious Pancreatic Necrosis (IPN)

- Infectious Pancreatic Necrosis (IPN) is a viral disease primarily affecting salmonid fish, including trout, salmon, and other related species. It is caused by the infectious Pancreatic necrosis virus (IPNV), a member of the Birnaviridae family (Munro *et.al.*, 2011)
- **Transmission:** IPN can be transmitted horizontally through direct contact between infected and susceptible fish or vertically from infected broodstock

to their offspring. It can also spread through contaminated water, equipment or infected fish feed.

- **Symptoms:** Infected fish may exhibit a range of clinical signs, including lethargy, skin darkening, abdominal swelling, and pale or hemorrhagic areas on the abdomen. Affected fish may also display erratic swimming behaviour, loss of appetite, and increased mortality rates.
- **Impact:** IPN can cause significant economic losses in aquaculture operations, particularly in salmonid farming. Mortality rates can be high, reducing production and financial implications for fish farmers.
- **Diagnosis:** Laboratory testing is used to confirm IPN infection in fish. Techniques such as virus isolation, PCR, and histopathological examination of affected tissues are employed to detect the presence of IPNV.
- **Prevention and Control**: Preventive measures for IPN include strict biosecurity protocols, proper disinfection of equipment and facilities, and minimizing the introduction and spread of the virus. Vaccination is available in some regions and can be used to reduce the impact of IPN in fish populations. It's worth noting that the specific regulations, surveillance programs, and management practices related to IPN can vary between countries and regions. Suppose you suspect the presence of Infectious Pancreatic Necrosis in a specific aquaculture context. In that case, it is recommended to consult local authorities, fish health professionals, or relevant regulatory bodies for specific guidance and information.

Bacterial Coldwater Disease (BCWD),

Bacterial Coldwater Disease (BCWD), also known as Coldwater Disease (CWD) or Rainbow Trout Fry Syndrome (RTFS), is a bacterial infection that primarily affects salmonid fish, especially trout and salmon. The bacterium *Flavobacterium psychrophilum* causes it.

- **Transmission:** BCWD can be transmitted horizontally through direct contact between infected and susceptible fish or vertically from infected broodstock to their offspring. The bacterium can also survive in water and on surfaces, facilitating its spread.
- **Symptoms:** Infected fish may exhibit various clinical signs, including darkening or reddening of the skin, lethargy, fin erosion, tail rot, exophthalmia (bulging eyes), and gill necrosis. Mortality rates can be significant, particularly in young fish. (Barnes *et. al.*, 2011)
- **Impact:** BCWD can cause substantial economic losses in aquaculture operations, particularly in cold-water environments where salmonids are farmed. It can result in high mortality rates, reduced growth, and increased susceptibility to other diseases.
- **Diagnosis:** Laboratory testing, including bacterial isolation and

identification, confirms BCWD in affected fish. Samples of affected tissues, such as the kidney or liver, are collected for analysis (Vallejo *et. al.*, 2014).

- **Prevention and Control:** Preventive measures for BCWD include strict biosecurity protocols, maintaining optimal water quality, minimizing stress on fish, and implementing vaccination programs where available. Antibiotic treatment may sometimes be used, although antimicrobial resistance can be a concern.
- It's important to note that the management and control strategies for BCWD can vary depending on the specific aquaculture context and regional regulations. Suppose there is suspicion of the presence of Bacterial Coldwater Disease in your aquaculture facility. In that case, consulting with local fish health experts, regulatory bodies, or fish health authorities is advisable for guidance and assistance in implementing appropriate measures.

Aeromonas infections

- Aeromonas infections, caused by bacteria belonging to the *Aeromonas* genus, are a group of diseases that can affect a wide range of fish species in aquaculture. Aeromonas bacteria are commonly found in aquatic environments and can cause various diseases in fish, including furunculosis and hemorrhagic septicemia.
- **Transmission:** Aeromonas infections can be transmitted through direct contact between infected and susceptible fish and waterborne routes. The bacteria can also enter fish through wounds or skin lesions.
- **Symptoms:** The clinical signs of *Aeromonas* infections can vary depending on the fish species and the specific strain of bacteria involved. Common symptoms include skin lesions, ulcers, fin rot, haemorrhaging, lethargy, loss of appetite and increased mortality rates (Janda *et.al.*, 2010).
- **Impact:** Aeromonas infections can have significant economic effects on aquaculture operations. In severe cases, they can cause high mortality rates, reduced growth, and increased susceptibility to other opportunistic infections (Igbinosa *et.al.*, 2012).
- **Diagnosis**: Laboratory testing is typically employed to confirm Aeromonas infections in fish. This may involve bacterial culture and identification and histopathological examination of affected tissues.
- **Prevention and control:** Preventive measures for Aeromonas infections include maintaining good water quality, implementing proper hygiene and disinfection protocols, and reducing stress on fish. Regular health monitoring and early detection of diseases can aid in implementing appropriate management strategies. Antibiotic treatment may be used in severe infections, but it should be done judiciously to minimize the risk of antimicrobial resistance.

- It's important to note that specific management practices and treatment options may vary depending on the regional regulations, the fish species and the infection's severity. Consulting with local fish health professionals, regulatory bodies, or aquatic veterinarians is recommended for specific guidance and assistance in managing Aeromonas infections in aquaculture settings.

Early Mortality Syndrome

- Early Mortality Syndrome (EMS), also known as Acute Hepatopancreatic Necrosis Disease (AHPND), is a bacterial infection that primarily affects shrimp, particularly the black tiger shrimp (*Penaeus monodon*) and the white leg shrimp (*Litopenaeus vannamei*) (Tran *et.al.*, 2013). EMS has emerged as a significant disease in shrimp aquaculture in recent years.
- **Causative Agent:** EMS is caused by a strain of the bacterium *Vibrio parahaemolyticus* that carries a specific virulence factor known as the PirAB toxin (Tran *et.al*, 2013). This toxin affects the hepatopancreas, which is the digestive organ in shrimp.
- **Transmission:** The disease can spread horizontally through direct contact between infected and susceptible shrimp and waterborne routes. It can also be introduced through infected post-larvae or contaminated water sources.
- **Symptoms:** Infected shrimp typically exhibit a sudden and significant increase in mortality rates, often occurring within the first 20 to 30 days of culture. Other symptoms include pale hepatopancreas, empty stomachs, lethargy, and abnormal feeding behaviour. Shrimp may also have an "empty gut" appearance (Tran *et.al.*, 2013).
- **Impact:** EMS has had severe economic consequences for shrimp aquaculture, leading to substantial losses in affected regions. Mortality rates can be extremely high, resulting in reduced yields and financial implications for shrimp farmers.
- **Diagnosis:** Laboratory testing is used to confirm EMS in shrimp. This may involve PCR-based detection of the PirAB toxin gene, histopathological examination of affected hepatopancreas tissues, or other specialized techniques.
- **Prevention and Control:** Preventive measures for EMS include strict biosecurity protocols, regular health monitoring of shrimp populations, and implementing measures to minimize the introduction and spread of the causative bacterium. Strategies such as pond disinfection, water treatment, and specific pathogen-free post-larvae can help reduce the risk of EMS outbreaks. There is ongoing research into developing resistant shrimp strains and effective vaccines.
- It's important to note that EMS management practices may vary between

countries and regions. Suppose you suspect the presence of Early Mortality Syndrome in your shrimp farm. In that case, consulting with local fish health experts, regulatory bodies, or shrimp health authorities is advisable for specific guidance and assistance in implementing appropriate measures.

Ichthyophthiriasis

- Ichthyophthiriasis, commonly called Ich, is a parasitic disease caused by the protozoan parasite *Ichthyophthirius multifiliis* (Dickerson *et.al.*, 2012). It affects many freshwater and marine fish species, causing significant health issues in aquaculture and aquarium settings.
- **Transmission:** *Ichthyophthirius multifiliis* has a complex life cycle involving a free-swimming stage (tomites) and an attached stage (trophonts). The parasite can be transmitted between fish through direct contact or waterborne routes. Tomites can also be present in the water, making it crucial to maintain proper hygiene and prevent contamination.
- **Symptoms:** Infected fish typically display small white spots or cysts on their skin, fins, and gills. These spots are caused by trophonts, which attach to the fish's body and feed on its tissue. Infected fish may show signs of irritation, such as rubbing against objects or flashing, and behavioural changes, such as loss of appetite, lethargy, and increased mucus production (Witeska *et.al.*, 2010).
- **Impact:** Ichthyophthiriasis can cause significant stress and mortality in fish populations. The attachment and feeding of trophonts on fish tissues can lead to tissue damage, compromised gill function, secondary infections, and even death, particularly in young or stressed fish.
- **Diagnosis:** The diagnosis of Ichthyophthiriasis is often based on the observation of white spots on fish, known as "ich." Microscopic examination of skin or gill samples can also be conducted to confirm the presence of *Ichthyophthirius multifiliis*.
- **Prevention and Control:** Prevention and control measures for Ichthyophthiriasis include maintaining good water quality, minimizing stress on fish, implementing quarantine protocols for new fish introductions, and ensuring proper hygiene and disinfection practices. Treatment options may include raising the water temperature, using chemical treatments (such as malachite green or formalin), or using medications specifically targeting the parasite. It's important to follow dosage instructions carefully and consider the species and size of fish being treated.
- Early detection and prompt treatment are crucial for managing Ichthyophthiriasis in aquaculture or aquarium settings. Consulting with fish health professionals, aquatic veterinarians, or experienced hobbyists can provide specific guidance and assistance in managing and treating Ichthyophthiriasis effectively.

Amoebic Gill Diseases

- Amoebic Gill Disease (AGD) is a parasitic infection primarily affecting farmed salmonids, including salmon and trout. It is caused by the amoeba *Neoparamoeba perurans*. AGD mainly targets fish gills and can lead to significant health issues in affected populations.
- **Transmission:** AGD can be transmitted horizontally through direct contact between infected and susceptible fish or waterborne routes. The amoeba can also exist in biofilms or sediment and infect fish through exposure to contaminated water.
- **Symptoms:** Infected fish may display various clinical signs, including increased mucus production, inflammation of the gills (gill hyperplasia), respiratory distress, reduced feeding, and lethargy. Severe cases can lead to gill damage, impaired oxygen exchange, and increased mortality rates (Botwright *et.al.*, 2021)
- **Impact:** AGD can have significant economic implications for salmonid aquaculture, particularly in marine net pen systems. Infected fish may experience reduced growth rates, increased susceptibility to other infections and mortality, leading to production losses.
- **Diagnosis:** Laboratory diagnosis of AGD involves microscopic examination of gill tissue samples. Histopathological examination or molecular techniques such as PCR may be used to detect the presence of *Neoparamoeba peruran* (Vega *et.al.*, 2014).
- **Prevention and Control:** Preventive measures for AGD include regular health monitoring of fish populations, maintaining optimal water quality, and implementing biosecurity protocols. Reducing stress on fish, such as minimizing handling and environmental stressors, can help prevent outbreaks. Treatment options may include bath treatments using hydrogen peroxide or other approved chemicals and freshwater baths to reduce amoeba numbers on fish gills temporarily.
- It's important to note that the management and control strategies for AGD may vary between countries and regions. Suppose you suspect the presence of Amoebic Gill Disease in your salmonid aquaculture operation. In that case, consulting with local fish health professionals, regulatory bodies, or aquatic veterinarians is advisable for specific guidance and assistance in managing and treating AGD effectively.

References

Abdelsalam, M., Elgendy, M. Y., Elfadadny, M. R., Ali, S. S., Sherif, A. H. and Abolghait, S. K. (2023). A review of molecular diagnoses of bacterial fish diseases. *Aquaculture International*, *31*(1), 417-434.

Austin, B. (2010).Vibrios as causal agents of zoonoses. *Veterinary microbiology*, *140*(3-4), 310-317.

Austin, J. H. (1999). *Zen and the brain: Toward an understanding of meditation and consciousness.*

Barnes, M. E. and Brown, M. L. (2011). A review of Flavobacterium psychrophilum biology, clinical signs, and bacterial cold-water disease prevention and treatment. *Open Fish Science Journal, 4*, 40.

Bernardet, J. F. and Grimont, P. A. (1989). Deoxyribonucleic acid relatedness and phenotypic characterization of *Flexibacter columnaris* sp. nov., nom. rev., *Flexibacter psychrophilus* sp. nov., nom. rev., and *Flexibacter maritimus Wakabayashi*, Hikida, and Masumura 1986. *International Journal of Systematic and Evolutionary Microbiology, 39*(3), 346-354.

Botwright, N. A., Mohamed, A. R., Slinger, J., Lima, P. C. and Wynne, J. W. (2021). Host-parasite interaction of Atlantic salmon (*Salmo salar*) and the ectoparasite *Neoparamoeba perurans* in amoebic gill disease. *Frontiers in immunology, 12*, 672700.

Branson, E. J. and Diaz Munoz, D. N. (1991). Description of a new disease condition occurring in farmed *coho salmon, Oncorhynchus kisutch* (Walbaum), in South America. *Journal of fish diseases, 14*(2), 147-156.

Bravo, S. and Campos, M. (1989). Coho salmon syndrome in Chile. *FHS/AFS Newsletter, 17*(3), 3.

Bruno, D. W. and Munro, A. L. S. (1986). Uniformity in the biochemical properties of *Renibacterium salmoninarum* isolates obtained from several sources. *FEMS Microbiology Letters, 33*(2-3), 247-250.

Bruno, M. and Easterly, W. (1998). Inflation crises and long-run growth. *Journal of Monetary economics, 41*(1), 3-26.

Chiller, K., Selkin, B. A. and Murakawa, G. J. (2001, December). Skin microflora and bacterial infections of the skin. In: *Journal of Investigative dermatology Symposium proceedings* (Vol. 6, No. 3, pp. 170-174).

Colley, D. G., Bustinduy, A. L., Secor, W. E. and King, C. H. (2014). Human schistosomiasis. *The Lancet, 383*(9936), 2253-2264.

Colquhoun, D. J. and Lillehaug, A. (2014). Vaccination against vibriosis. *Fish vaccination*, 172-184.

Datta, A. and Prescott, L. M. (1969). Effect of vibriocins on members of the Enterobacteriaceae. *Journal of Bacteriology, 98*(2), 849-850.

Denkin, S. M. and Nelson, D. R. (2004). Regulation of *Vibrio anguillarum* empA metalloprotease expression and its role in virulence. *Applied and environmental microbiology, 70*(7), 4193-4204.

Dickerson, H. W. (2012). *Ichthyophthirius multifiliis*. In *Fish parasites: pathobiology and protection*, 55-72. Wallingford UK: CABI.

Eissa, A. and Elsayed, E. E. (2006). An overview on bacterial kidney disease. *Life Sciences Journal, 3*(3), 58-76.

Esteve-Gassent, M. D., Fouz, B. and Amaro, C. (2004). Efficacy of a bivalent vaccine against eel diseases caused by *Vibrio vulnificus* after its administration by four different routes. *Fish and shellfish immunology, 16*(2), 93-105.

Hjerde, E., Karlsen, C., Sørum, H., Parkhill, J., Willassen, N. P. and Thomson, N. R. (2015). Co-cultivation and transcriptome sequencing of two co-existing fish pathogens *Moritella viscosa* and *Aliivibrio wodanis*. *BMC genomics, 16*(1), 1-13.

Igbinosa, I. H., Igumbor, E. U., Aghdasi, F., Tom, M. and Okoh, A. I. (2012). Emerging *Aeromonas* species infections and their significance in public health. *The Scientific World Journal, 2012*.

ISIDAN, H. and Bolat, Y. (2011). A survey of viral hemorrhagic septicemia (VHS) in Turkey. *Turkish Journal of Fisheries and Aquatic Sciences, 11*(4).

Janda, J. M. and Abbott, S. L. (2010). The genus Aeromonas: taxonomy, pathogenicity, and infection. *Clinical microbiology reviews, 23*(1), 35-73.

Le Breton-Miller, I. and Miller, D. (2009). Agency vs. stewardship in public family firms: A social embeddedness reconciliation. *Entrepreneurship theory and practice, 33*(6), 1169-1191.

Lyngstad, T. M., Qviller, L., Sindre, H., Brun, E. and Kristoffersen, A. B. (2018). Risk factors associated with outbreaks of infectious salmon anemia (ISA) with unknown source of infection in Norway. *Frontiers in veterinary science, 5*, 308.

Moreira, M., Schrama, D., Farinha, A. P., Cerqueira, M., Raposo de Magalhaes, C., Carrilho, R. and Rodrigues, P. (2021). Fish pathology research and diagnosis in aquaculture of farmed fish; a proteomics perspective. *Animals, 11*(1), 125.

Munro, E. S. and Midtlyng, P. J. (2011). Infectious pancreatic necrosis and associated aquatic birnaviruses. In *Fish diseases and disorders. Volume 3: viral, bacterial and fungal infections* (pp. 1-65). Wallingford UK: CABI.

Richards, C.A., Murphy, C.A., Brenden, T.O., Loch, T.P. and Faisal, M. (2017). Detection accuracy of *Renibacterium salmoninarum* in Chinook salmon, *Oncorhynchus tshawytscha* (Walbaum) from non-lethally collected samples: Effects of exposure route and disease severity. *Preventive Veterinary Medicine, 145*, 110–120

Romalde, J. L. (2014). Vaccination against photobacteriosis. *Fish Vaccination*, 200-210.

Romalde, J. L. and Toranzo, A. E. (2002). Molecular approaches for the study and diagnosis of salmonid streptococcosis. *Molecular diagnosis of salmonid diseases*, 211-233.

Toranzo, A. E. and Barja, J. L. (1993). Virulence factors of bacteria pathogenic for coldwater fish. *Annual Review of Fish Diseases, 3*, 5-36.

Toranzo, A. E., Magariños, B. and Romalde, J. L. (2005). A review of the main bacterial fish diseases in mariculture systems. *Aquaculture, 246*(1-4), 37-61.

Toranzo, A. E., Magariños, B. and Romalde, J. L. (2005). A review of the main bacterial fish diseases in mariculture systems. *Aquaculture, 246*(1-4), 37-61.

Tran, L., Nunan, L., Redman, R. M., Mohney, L. L., Pantoja, C. R., Fitzsimmons, K. and Lightner, D. V. (2013). Determination of the infectious nature of the agent of acute hepatopancreatic necrosis syndrome affecting penaeid shrimp. *Diseases of aquatic organisms, 105*(1), 45-55.

Vallejo, R. L., Palti, Y., Liu, S., Marancik, D. P. and Wiens, G. D. (2014). Validation of linked QTL for bacterial cold-water disease resistance and spleen size on rainbow trout chromosome Omy19. *Aquaculture, 432*, 139-143.

Varina, M., Denkin, S. M., Staroscik, A. M. and Nelson, D. R. (2008). Identification and characterization of Epp, the secreted processing protease for the *Vibrio anguillarum* EmpA metalloprotease. *Journal of bacteriology, 190*(20), 6589-6597.

Valdenegro Vega, V. A. C. (2014). Using the mucosal response to recombinant *Neoparamoeba perurans* attachment proteins to design an experimental vaccine against amoebic gill disease (AGD). (Doctoral dissertation, University Of Tasmania).

Verjee, M. A. (2019). Schistosomiasis: still a cause of significant morbidity and mortality. *Research and reports in tropical medicine*, 153-163.

Vinay, T. N., Kim, Y. J., Jung, M. H., Kim, W. S., Kim, D. H. and Jung, S. J. (2013). Inactivated vaccine against viral hemorrhagic septicemia (VHS) emulsified with squalene and aluminum hydroxide adjuvant provides long term protection in olive flounder (*Paralichthys olivaceus*). *Vaccine, 31*(41), 4603-4610.

Wakabayashi, S., Matsubara, H. and Webster, D. A. (1986). Primary sequence of a dimeric bacterial haemoglobin from Vitreoscilla. *Nature, 322*(6078), 481-483.

Witeska, M., Kondera, E. and Lugowska, K. (2010). The effects of Ichthyophthiriasis on some haematological parameters in common carp. *Turkish Journal of Veterinary and Animal Sciences, 34*(3), 267-271.

World Health Organization (2000). WHO global principles for the containment of antimicrobial resistance in animals intended for food: report of a WHO consultation with the participation of the Food and Agriculture Organization of the United Nations and the Office International des Epizooties, Geneva, Switzerland 5-9 June 2000 (No. WHO/CDS/CSR/APH/2000.4). World Health Organization.

INDEX

T

V

W

Z